National Transport Infrastructure Conference with International Participation (NTICIP 2023)

IOP Conference Series: Materials Science and Engineering
Volume 1297

Nessebar, Bulgaria
5 - 7 October 2023

Editors:

Andrey Totsev
Petar Nikolov
Lazar Georgiev

ISBN: 978-1-7138-8734-8
ISSN: 1757-8981

Printed from e-media with permission by:

Curran Associates, Inc.
57 Morehouse Lane
Red Hook, NY 12571

Some format issues inherent in the e-media version may also appear in this print version.

Institute of Physics
Dirac House, Temple Back
Bristol BS1 6BE UK

Phone: 44 1 17 929 7481
Fax: 44 1 17 920 0979

techtracking@iop.org

Additional copies of this publication are available from:

Curran Associates, Inc.
57 Morehouse Lane
Red Hook, NY 12571 USA
Phone: 845-758-0400
Fax: 845-758-2633
Email: curran@proceedings.com
Web: www.proceedings.com

TABLE OF CONTENTS

IOP Conf. Series: Materials Science and Engineering 1297 (2023) 011001 doi:10.1088/1757-899X/1297/1/011001

FACULTY OF TRANSPORTATION ENGINEERING — INSTITUTE OF TRANSPORTATION ENGINEERING — Bulgaria PIARC

XVI NATIONAL TRANSPORT INFRASTRUCTURE CONFERENCE WITH INTERNATIONAL PARTICIPATION

The transport infrastructure is an issue that concerns every member of our society regardless of age, gender, or race. The necessity to discuss its flaws and make amends to do better each day is constant and undeniable, but with the rapid development of computer technology now, more than ever, we must be flexible and adapt to the innovations in our sector.

During the **XVI National Transport Infrastructure Conference with international participation (NTICIP 2023)** we celebrated and shared our achievements, but also paid attention to the problems and challenges ahead of us. Our main goal as an organizer is to create a sustainable connection between **administration, science, and business**.

This year (2023) the main emphasis was on **Road Safety, Digitalization, Technology, Strategic Infrastructure, Standards, Geotechnical Engineering and Sustainable Materials.**

Each guest who took part in the Conference, held from 5th to 7th of October 2023 in Nessebar, Bulgaria, had the opportunity to discuss the scientific side of engineering issues, exchange information and develop professional contacts.

Organizing one of the biggest professional forums on the Balkan Peninsula is not an easy task. I am pleased to acknowledge the work of my colleagues at the Institute of Transportation Engineering and the University of Architecture Civil Engineering and Geodesy. But most of all we are all grateful to the authors for entrusting us with their work.

Prof. Dr. Eng. Nikolay Mihaylov,
Chairman of scientific board at Institute of transportation engineering

IOP Conf. Series: Materials Science and Engineering 1297 (2023) 011001 doi:10.1088/1757-899X/1297/1/011001

IOP Conf. Series: Materials Science and Engineering 1297 (2023) 011002 doi:10.1088/1757-899X/1297/1/011002

Peer Review Statement

All papers published in this volume have been reviewed through processes administered by the Editors. Reviews were conducted by expert referees to the professional and scientific standards expected of a proceedings journal published by IOP Publishing.

1. Type of peer review: Single anonymous
2. Conference submission management system: Morressier
3. Number of submissions received: 27
4. Number of submissions sent for review: 27
5. Number of submissions accepted: 22
6. Acceptance Rate (Submissions Accepted / Submissions Received × 100): 81.48 %
7. Average number of reviews per paper: 1.00
8. Total number of reviewers involved: 12
9. Contact person for queries:
   Name: Eng. Georgia Kineva
   Affiliation: Institute of transportation engineering Ltd
   Email: conference@institute-tsi.com

IOP Conf. Series: Materials Science and Engineering 1297 (2023) 012001 doi:10.1088/1757-899X/1297/1/012001

Drainage of road bridges

Peter Nikolov

Associate Prof. Dr. Eng., "Roads and Transport Structural Facilities Department", Faculty of Transportation Engineering, UACEG, Sofia 1046 Hristo Smirnenski 1 blvd, Republic of Bulgaria

e-mail: nikolov_fte@uacg.bg

Abstract. When designing bridges, the main focus is usually on the design of structural elements. However, some important issues related to the safe operation and maintenance of these facilities are sometimes ignored. One such issue is related to the drainage of bridge roadways. In some countries with a well-working system of inspection and maintenance of road networks and of bridges in particular, there are clear rules for the design of bridge drainage. Unlike the common practice in Bulgaria, the goal of bridge drainage design in these countries is to minimize the number of scuppers and to look for appropriate ways to minimize the amount of water in the bridge area.In this paper, the available requirements of Bulgarian design documents and regulations related to drainage systems are studied and a comparison with the bridge design manuals of some US transportation administrations is made

Key words: Road bridges, drainage, maintenance

1. Introduction

Reliable drainage of water from roadways is essential to ensure the safe traffic on roads. This also applies to bridges, where the amount of water is usually greater, since in addition to the roadway, the sidewalks must also be drained. Due to the peculiarities of the climate in Bulgaria, during the winter months the roads are treated with anti-icing agents, which makes drained water highly aggressive to concrete, structural and reinforcing steel, which are the building materials predominantly used for road bridges.

In the case of poor drainage solutions - both the availability and location of the scuppers, as well as the detailing of their design and implementation, and the relatively low level of maintenance, significant damage to the elements of the main structure was found in a number of bridges (Figure 1).

Figure 1. Damage to the main prestressed bridge beam in the area of the scuppers

If adequate measures are not taken to eliminate the causes of water leaks and to repair the affected elements, after a comparatively short period of operation, the damage becomes significant and the bearing capacity of the bridge structure is reduced, which in some cases is the reason for the pre-emergency conditions of important road network facilities (Figure.2).

Figure 2. Damage to high-strength reinforcement in a prestressed bridge beam in the area of the scuppers

2. Reqirements of the Bulgarian design regulations

The Bulgarian codes and standards for bridge design from the distant and recent past, contain few guidelines and requirements for roadway drainage design.

In "Temporary Regulations for the Design of Concrete and Reinforced Concrete Road Bridges and Culverts", an excellent document in terms of structure and content from 1973 [1], there is only one article that deals with bridge drainage and it states:

"For the removal of surface water on the bridges, scuppers are placed over a distance no greater than 20 m. The scuppers must ensure quick drainage of the water that has been discharged in them. They must allow for easy cleaning and prevent runoff water from splashing bridge parts in windy conditions. Overpass drains are not allowed to discharge water directly onto the bridged road."

Sixteen years later, in 1989, in the "Code for the Design of Road and Railway Bridges and Culverts" [2] there is already a separate chapter containing five articles that give more detailed requirements, but in relation to road bridges there are only two, namely that the number and location of scuppers is determined depending on the volume of runoff and thelocation of the bridge, and that in case of inflow of surface water from the approaches to the bridge, drainage devices must be provided. Again, the maximum distance between scuppers is fixed at 20 m.

In compliance with the described requirements, all bridges designed and built after the 1970s and 1980s have a relatively large number of scuppers regardless of the amount of rainfall intensity and the expected water flow. An example of this is the Bebresh viaduct along the Hemus highway, where the scuppers are at a distance of about 15 m (Figure 3). With a height of the piers of up to 120 m, almost constant winds and lack of a collector system, the requirements for not splashing of parts of the bridge cannot be complied with in practice.

Figure 3. The Bebresh viaduct along the Hemus highway

In 1997, "Instructions for the Application and Technical Specifications for Scuppers on Road Bridges and Facilities" [3] were issued, in which the requirements for the design of the drainage of the facilities are regulated in much more detail. These instructions contain rules for determining design water quantities and maximum distances between scuppers. It is accepted that the maximum puddle width is 150 cm, measured from the curb to the axis of the road, and the maximum height of the puddle near the curb should not exceed 5 cm (Figure 4).

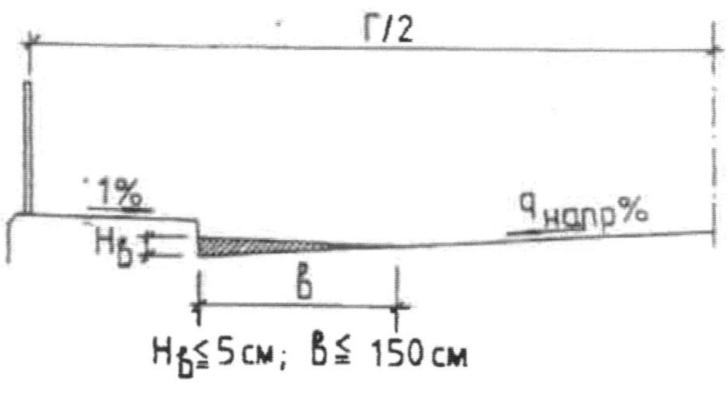

Figure 4. Scheme for determining the hydraulic characteristics of the flow [3]

The design rainfall has a duration of 10 minutes with a return period of one year. Taking into account the possible clogging and the reduction of the capacity of the grids, the drained area for a signle scupper is limited to 380m2. For highway bridges of a width of about 15 m (total for the lanes and sidewalks), the maximum distance between scuppers is set at 25 m – 5 m larger compared to the requirements of [1] and [2].

It should be borne in mind that the instructions of [3] refer to the roads of the Bulgarian Republican Road Network only, which are managed by the Bulgarian Road Infrastructure Agency, and are not mandatory for other bridges – those on municipal roads, streets, etc.

3. Requirements of certain US transport administrations

When designing a bridge in the USA, in addition to the national normative documents, the so-called 'bridge design manuals' that each State Transportation Administration prepares and issues and that apply to the territory of the respective state are used. These manuals have a volume of about 1600 - 2000 pages each and give the requirements of the employer regarding a number of issues, some of which are not affected by the normative documents for the design of bridge structures.

The Design Guides of three state transportation administrations - the states of New York [4], New Jersey [5] and Wisconsin [6] - are examined.

In all three documents, the general trend is towards minimizing the number of scuppers. In [4], it is recommended that there are no scuppers in facilities with a length of less than 100 m. In [5] and [6] this limit is 120 m. Under special conditions, for example, wide bridges, small longitudinal slopes, sections with transitional curves, it is necessary to provide drains for smaller lengths.

Unlike the Bulgarian practice, the maximum distance between the scuppers in [6] is 150 m, which is 7.5 times greater than the 20 m of [1] [2], and six times greater than the 25 m of [3].

In all three manuals, great attention is paid to additional measures to improve the drainage of the roadway in the bridge area, such as:

- ensuring reliable removal of surface waters from the approaches of the bridge so that these waters do not enter the facility;
- additional scuppers to cut off the flow of water to the expansion joints, thereby minimizing damage to the structure caused by leaks through the joints;
- placement of scuppers away from transverse beams.

In all cases, a detailed design of the drainage of each bridge is required based on the design water quantities and the characteristic features of each facility.

Analogously to the formation of puddles on the roadway (Fig.4) adopted in [3], in [4], for design speeds below 75 km/h, puddles are allowed to be of such a width that a section of at least 2.5 m of the roadway remains unflooded, and the maximum width of the spread of water is 3.60 m. In case of two 3.5-metre traffic lanes, this results in a wider spread area, compared to the 150 cm limit fixed in [3] (Figure 5).

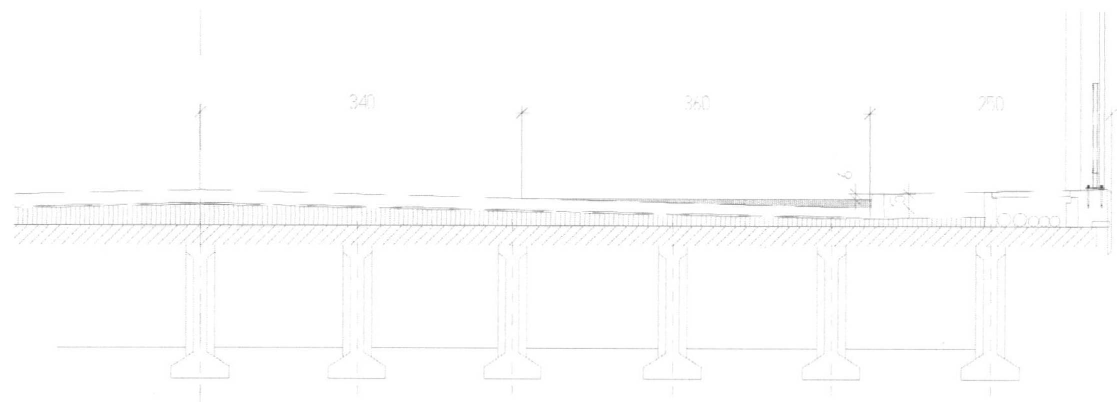

Figure 5. Permissible spread width of the roadway subject to the requirements of [4] for design speeds below 75 km/h

For highways with design speeds equal to or greater than 75 km/h, puddles should not encroach into any portion of a travel lane.

4. Conclusion

Drainage is essential to the safe operation and longevity of bridges. The presence of more than the required number of culverts and their placement in inappropriate places, combined with a low level of maintenance, can lead to the pre-emergency conditions of bridge structures and/or to very expensive repairs (Figure 6).

Figure 6. Replaced main beam of the bridge from Figure 1

It is necessary to prepare a Bulgarian specific document for the design of drainage of road facilities, which could also be a part of a complete regulation for the design of bridges in Bulgaria, the need for which is urgent [7].

When defining the rules, both the Bulgarian experience (both good and bad practices) and the practice of other countries and administrations should be taken into account.

References

[1] Temporary regulations for design of concrete and reinforced concrete road bridges. Main Directorate of Roads to the Ministry of Transport. State publishing house "Technique" Sofia 1973. (In Bulgarian)

[2] Regulations for design of road and railway bridges and culverts. Committee on Territorial and Settlement Planning. Ministry of Transport. Sofia 1989. (In Bulgarian)

[3] Application guidelines and technical specification for road bridge scuppers. Main Directorate of Roads to the Ministry of Transport. Sofia 1997г. (In Bulgarian)

[4] Bridge manual, New York State Department of transportation 2021

[5] Design manual for bridges and structures, State of New Jersey. Department of transportation. 6-th edition. 2016

[6] Bridge manual, State of Wisconcsin. Department of transportation. January 2019.

[7] Nikolov. P. Guidelines for the content of the Regulations for the design of bridges in Bulgaria. Annual of UACEG ISSN 2354-9759, Volume 52 / 2019, Issue 3, pp.(739-752), Sofia. (In Bulgarian)

Study of Problems Related to Pedestrian Bridges in Bulgaria

Lazar Georgiev

Associate Prof. Dr. Eng., "Roads and Transport Structural Facilities Department", Faculty of Transportation Engineering, UACEG, Sofia 1046 Hristo Smirnenski 1 blvd, Republic of Bulgaria

lazar_fte@uacg.bg

Abstract. Pedestrian bridges due to specific features of the traffic on them have different parameters of structural behaviour and proportions in comparison with road and railway bridges. The lower traffic intensity leads to the possibility of realisation of slender structures with complex geometry. In accordance with Bulgarian national regulations the importance category of the bridge structures is related only with the road passing or under the structure. In Bulgaria there are large number of pedestrian bridges with relatively large spans up to 260m on small village roads and forest roads. Due to their low category in accordance with national regulations the process of control in their design, execution and maintenance is realised by the relevant municipalities. In the case of large spans suspension and cable- stayed superstructures are often used but the technical staff in the municipalities is frequently with no enough expert level in the complex behaviour of such type of structures. In the paper some common structural deficiencies, illustrated on analysis of particular pedestrian bridges in exploitation are discussed.

Key words: *Pedestrian bridges, regulation, structural deficiencies, analysis, maintenance*

1. Introduction

Pedestrian bridges due to specific features of the traffic on them have different parameters of structural behavior and proportions in comparison with road and railway bridges. The lower traffic intensity leads to the possibility of realization of slender structures with complex geometry, also innovative materials can be implemented [12]. In pedestrian bridges with large spans often there are problems with horizontal stiffness due to relatively low width needed for pedestrian traffic. The concept of design standards is focused on definition of prescriptions which ensures an accepted reliability level. In Eurocode the reliability level is evaluated by means of the reliability index β, connected with the probability of failure P_f.

$$P_f = \Phi(-\beta) \quad \text{(f.1)}$$

For normal distribution random variables it is well known that

$$\Phi(x) = \frac{1}{\sqrt{2\pi}} \int_{-\infty}^{x} e^{\frac{t^2}{2}} dt \quad \text{(f.2)}$$

For the prescribed in [1] for ULS reliability index 3,8 for load bearing capacity the probability of failure of the relevant structural member is

$$P_f = \Phi(-3{,}8) = 7{,}235.10^{-5} \quad \text{(f.3)}$$

In accordance with Bulgarian national regulations the importance level of bridges is connected only with the road/railway category and in some cases footbridges with large spans and relatively complex superstructure are classified in the low category 4 (1 is the highest and 6 is lowest) when they are along/over low category roads (agricultural roads, Eco paths etc.). For large spans suspension and cable stayed bridges are often used. These are superstructure with complex nonlinear behavior and the combination of insufficient experience and/or responsibility of designers, builders and authorities (the relevant municipalities) leads in many cases to realization of bridges with serious deviations from the design requirements and in some cases dangerous for exploitation.

In the world practice pedestrian bridges with relatively large spans are realized and up to now the largest span is 721m of the SKY BRIDGE 721 in Czech Republic opened for traffic in 2022 [3] – Figure 1.1 [11]. Such type of bridges are much complex for design, execution and exploitation, so high level of responsibility and competence are needed from designers, builders and authorities controlling the design and execution processes.

Figure 1.1. SKY BRIDGE 721, [11]

Nevertheless that the intensity of live load for footbridges is much lower in comparison with road and railway bridges, neglecting some specific aspects in their behavior can lead in many cases to serious construction deficiencies and in some cases tragic consequences. In 2022 suspension bridge in India failed after reconstruction due to overloading (several hundred passengers passing the bridge) with over 130 fatalities [10]. In Bulgaria the failure of the Lisicite bridge due to snow load in 2000 is connected with one fatality and the failure of a suspension bridge over Tundja river in Elhovo in 1996 is connected with 11 fatalities.

In Bulgaria there are footbridges with span up to 260m. On Figure 1.2 the pedestrian bridge for connection of the Lisicite village with Kardjali – Podkova railway line over the "Studen Kladenec" artificial lake with 260m span is presented. In the 2000 there was failure of the superstructure with one fatality under snow load, which is much lower in comparison with the 2,5kN/m^2 live load prescribed in the design standard Eurocode [7]. After reconstruction there is low horizontal and torsional stiffnes, the main cable do not fulfil the requirements of [8] for minimal bending radius for the strands forming the cable, the anchor details are in zone of variable water level – Figure 1.3. The detail of hangers is unreliable for fixing the main cable in longitudinal direction. Also part of hangars are spiral ropes, welded to steel parts, which is a detail with unclear load bearing capacity and with expected very poor behavior concerning brittle fracture – Figure 1.4.

Considering described serious deviations from design requirements and good practices the probability of failure for particular components of the bridge is expected to be much higher than the target probability in Eurocode defined by relevant reliability index.

Figure 1.2. Lisicite footbridge with L=260m

Figure 1.3. Anchoring the main cable strands

Figure 1.4. Detail of a hangar to deck connection

The main cable of a suspension bridge and its anchorage are critical elements of great importance for the overall load bearing capacity of the structure. As it is mentioned above the low level of competency, responsibility and control can lead to bridge structures with serious structural deficiencies and dangerous for exploitation.

Figure 1.5. Anchoring of the main cable of suspension bridge over Iskar River

Figure 1.6. Main cable of suspension bridge near Podkova Rail station

On Figure 1.5 anchor detail of the main cable of a suspension bridge with 67m span over the Iskar River can be seen. It is a craft made detail with unclear load bearing capacity – the main cable is anchored by means of steel tube (failed now) filled with cement/concrete. Compromising of this critical detail can lead to total collapse of the bridge superstructure.

On Figure 1.6 the torsional deformation of main cable consisting 12 strands of suspension bridge with 260m span near Podkova Railway station can be seen. The lack of adequate horizontal and torsional stiffness of the deck is the reason that bridge is unusable during strong wind and defects on the structure can occur is such situation.

2. Case studies concerning condition of footbridges in Bulgaria.

In [4] several long-span footbridges over the Varbitca River in Rodopi Mountains connecting remote villages with the 4 main railroad line Ruse-Kardjali-Podkova. This line is a part of the intensive development of the railway infrastructure in Bulgaria with serious modernization potential [5], [6]. Pedestrian bridges with structural deficiencies in Rodopi montain are also described in [9]. Also construction solutions under discussion for footbridges with small and medium spans exist in a number of places, again due to a combination of the factors mentioned above.

Described below bridges are representative for illustration of some general problems related with structural deficiencies and relatively poor detailed design of different types of footbridges in Bulgaria.

2.1. Pedestrian – foot bridge over Varbitsa River near Erovete village in Bulgaria

The Figure 2.1.1 shows a suspension pedestrian bridge with span 117,5m over the Varbitsa river near the village of Erovete connecting the village to the main road between Kardjali and Zlatograd. The bridge is built in 1971 according to citizens of the village. The width of the stiffening girder is 2230mm and this allows light vehicle passing through the bridge. The main cable of the bridge is constructed from 12x14mm strands of open type ropes from high strength steel wires which is not acceptable solution without additional corrosion protection and will reduce the service life of this critical component.

Figure 2.1.1. General view of the bridge

Figure 2.1.2. Bridge deck

IOP Conf. Series: Materials Science and Engineering 1297 (2023) 012002 doi:10.1088/1757-899X/1297/1/012002

Figure 2.1.3. General view of the bridge

Figure 2.1.4. Repaired hanger

The deck is reinforced concrete plate poured on steel sheet. The plate is supported by longitudinal and cross girders. The cross girders are trusses with reinforcing bar elements with welded connection and without supports of lower flange in longitudinal direction – Figure 2.1.2. This construction do not fulfill the requirements of design standards and its load bearing capacity is hard to evaluate. As it shows Figure 2.1.3 the anchor detail of the ropes forming the main cable the bending radius of the ropes is low and the clamping detail is unreliable so the bearing capacity of the critical component – main cable cannot be evaluated precisely. Fig 2.1.4 shows failed hanger and local failure of the upper cord of the stiffening girder, repaired by craft made additional components with load bearing capacity hard to evaluate. In conclusion nevertheless the relatively large span of this bridge there are construction deficiencies due to poor design, execution and control and these deficiencies will reduce significantly the expected service life and increase the probability of failure of relevant bridge components.

2.2. Pedestrian bridge in Sofia over Vladaiska River

Footbridge with 21m span over the Vladaiska River is presented of Figure2.2.1 (several bridges are build over the river in Sofia). The bridge superstructure is single span simply supported steel truss with 2T section of the lower cord and closed rectangular hot rolled steel sections of upper cord, diagonals and verticals. The cross girders of the deck are with 2T cross section – Figuere 2.2.2.

Figure 2.2.1. Bridge superstructure

Figure 2.2.2. Bridge deck

The connection between the lower cord of the main truss and the cross girders is realized by welds between contacting lower flange of lower cord of the truss and upper flange of the cross girder. This detail does not ensure rigid connection and its structural behavior is complex and unreliable. There is no horizontal bracing at the level of the lower cord of the main truss.

Figure 2.2.3. Connection detail at midspan

Figure 2.2.4. Upper bracing

The deck consists of steel sheet and reinforced concrete plate. Steel sheet is connected to cross girders by means of self-tapping screws and as it can be seen there is leakage and corrosion around drilled openings. The bracing at the level of the upper cord is presented on Figure 2.2.4. The horizontal stiffness of the bracing varies serious between fields and the overall effectiveness is doubtful. The connection between assembly elements of main truss is at the midspan with bolted end-plate joint – Figure 2.2.3. The rotational stiffness of this detail is important concerning deformation and dynamic behavior of the main girder. The lack of effective horizontal bracing and the bolted end-plate joints leads to relatively high vibration amplitudes from passing only several pedestrians nevertheless the truss is assumed as rigid variant of main girder of a bridge. This example illustrates that in the case of suitable choice of the main girder type, poor design of subsystems and details can lead to unacceptable behavior and reducing exploitation period.

3. Conclusion

In the conceptual design concepts of Robustness must be applied when it is appropriate.

Detailed design, execution and control of pedestrian bridges should be carried out competently and responsibly in compliance with the requirements for the provision of the required quality.

The management of pedestrian bridges should ensure adequate maintenance and adequate actions when defects and/or problems are registered.

A discussion is needed regarding the adequacy of the regulatory framework regarding classifying footbridges not only by importance of the road, but including specifics and complexity of the structural behavior of the particular bridge.

There are serious problems with providing enough competent and responsible personnel at all levels in the processes of design, construction, control and operation of the bridges, which is related to the lack of a sufficient level of awareness of the importance of the engineering profession in society and hence of its prestige.

Currently, due to the listed reasons of a complex nature, the level of reliability of a significant number of pedestrian bridges in our country is unsatisfactory and in case that no changes in the opposite direction are realized, it is logical to expect in the near future the probability of reaching accidents to increase.

Acknowledgments

The author would like to express his acknowledgements to the Research, Consultancy and Design Centre (RCDC) of the UACEG, Sofia, Bulgaria for financing the contract BN-281-23.

IOP Conf. Series: Materials Science and Engineering 1297 (2023) 012002 doi:10.1088/1757-899X/1297/1/012002

References

[1] BDS EN 1990:2003, Eurocode - Basis of structural design
[2] REGULATION NO. 1 OF JULY 30, 2003 ON THE NOMENCLATURE OF TYPES OF CONSTRUCTION, 2017, Republic of Bulgaria.
[3] E-mosty, International Online Peer-Reviewed Magazine About Bridges, 3/2022, ISSN: E-MOSTY 2336-8179, www.e-mosty.cz
[4] Stoyan Ivanov, Lazar Georgiev, "FOOTBRIDGES OVER VARBITZA RIVER", LONG SPAN BRIDGES 26-28 APRIL 2023, IABSE Symposium Istanbul 2023, pp.944-951, ISBN 978-3-85748-191-8
[5] Lepoev M. and Tasev Y., "Present concepts for development of Bulgarian Railways", Geopolitika i Geostrategia, Issue 6/2016, ISSN 1312-4579, pp.122-129 (in Bulgarian).
[6] Tasev Y. And Lepoev M., "For accelerated modernisation of the railway lines", Jelezopaten Transport, Issue 9/2009, ISSN 1310-683X, pp.12–15 (in Bulgarian).
[7] Eurocode 1 - Actions on structures - Part 2: Traffic loads on bridges and other civil engineering works, prBDS EN 1991-2:2023
[8] Eurocode 3 - Design of steel structures - Part 1-11: Tension components, prBDS EN 1993-1-11:2020
[9] Dulevski E. and Ivanov St., "Assessment of the condition of pedestrian suspension bridges in the Rodopi Mountain", Third National Conference on "Design and Construction of Buildings and Facilities", Plovdiv, 2004 (in Bulgarian)
[10] https://www.bbc.com/news/world-asia-india-63445154, 2023
[11] https://en.wikipedia.org/wiki/Sky_Bridge_721#/media/File:Sky_Bridge_721_-_spring.jpg, 2023
[12] Ivanov, S., Rajasingam, J., Mack, T. Weber, M., Hejda, N., Kagraner Steg - New GFRP Pedestrian Bridge in Vienna, IABSE Symposium Prague 2022: Challenges for Existing and Oncoming Structures - Report, 2022, pp. 1468–1475, ISBN 9781713853220

Testing of a temporary railway bridge and analysis of the results

Ivanka Topurova

University of Architecture, Civil Engineering and Geodesy, Sofia, Bulgaria

e-mail : ivankatopurova@yahoo.com

Abstract. The main purpose of bridge testing, an obligatory step before putting a railway bridge in operation, is to ensure that the used theoretical model is in accordance with the actual behaviour of the structure. Furthermore, bridge testing can point to some problems or inaccuracies due to construction.

Even though temporary bridges typically have shorter spans, they have relatively low self-weight and the main strains are caused by live loads. Therefore, bridge testing is particularly important to ensure the proper performance of such structures.

This paper outlines some major steps of the testing process and draws conclusions based on a temporary railway bridge located along the Kalotina Highway in Bulgaria.

1. Introduction

This paper focuses on the testing process and test results of a temporary railway bridge located along the Kalotina Highway in South-West Bulgaria in order to highlight the necessity of such procedures for temporary structures as well.

The purpose of bridge testing is to analyse a newly constructed bridge as compared to its design. In-situ testing can determine whether the construction meets the design requirements and whether it will provide safety during operation. In addition, bridge testing aims to experimentally compare the parameters accepted in the theoretical model with the actual values determined on site, which is necessary for the preparation of the final report, an obligatory element of the commissioning procedure.

2. Bridge case study

The investigated temporary bridge structure spans a temporary road servicing the construction works of the newly-built Kalotina Highway [1], [2]. The bridge consists of two simply supported beams with spans of 8.70m each. In the area of the structure, the railway track is designed along a horizontal curve with a radius of 190m and a longitudinal gradient of -20.765‰.

The open-type superstructure is made entirely of steel and consists of two main HE600B beams spaced at a distance of 150cm from each other, connected by 4 crossbeams of the same profile - one at each end and two dividing the span in three equal parts. Horizontal X-type bracings with box-shaped cross sections are placed at the lower and upper flanges. In the area of the temporary facility, the railroad is laid on wooden sleepers.

Figure 1. Case study temporary bridge

Taking into account the relatively short period of operation of the bridge, the connections between the superstructure and the substructure, both at the abutments and at the pier, is designed with elastomeric bearings type B 200/250/52mm glued to the main beams and to the concrete substructure by means of an adhesive composition. Since the bridge is temporary, instead of LM71 [3] load model, a load representing "the heaviest real loads in the network - one axle of 23 tons and a diesel locomotive series 06 with 6 axles of 19.4 tons each" was implemented, according the approach adopted in [4]. Buffers were designed in order to absorb the horizontal forces.

Based on the theoretical model, the maximum deflection obtained in the middle of the span due to moving loads was 9mm or 1/1035 of the static span.

3. In situ testing

The main steps of the testing process and the main parameters adopted are listed below.

3.1. Load

According to the requirements of [5], the vehicles to be used in the non-electrified sections are locomotives Series 06 (with six axles of 194kN) and locomotives Series 07 (with six axles of 200kN). However, due to the relatively short span of the main beams, the bridge testing was carried out with hopper wagons with a total mass of about 88t. As this was measured in situ, there was a difference of the distribution of the axial loads of the hopper in the transverse direction. The deviation in weights made it necessary to carry out additional calculations of the theoretical values of the measured deflections. In these calculations, the established proportional load of the wagon in the transverse direction was taken into account - the overload on one side was about 30%.

3.2. Theoretical model

A 3D computational model of the structure was created in order to obtain the values of the studied parameters. The different elements were modelled through shell and frame elements with realistic cross sections and an elasticity modulus of 210 GPa steel.

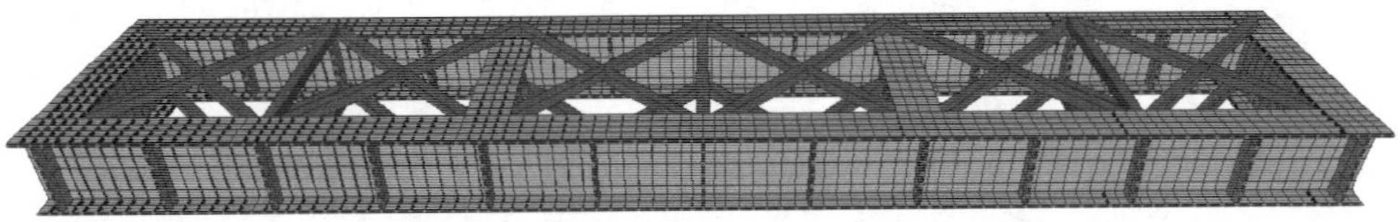

Figure 2. 3D model of the structure

3.3. Load cases

In accordance with the preliminary report, two load states were carried out - one for each bridge span. In both cases, the situation of the wagons was identical – both axles were placed symmetrically in the middle of the span.

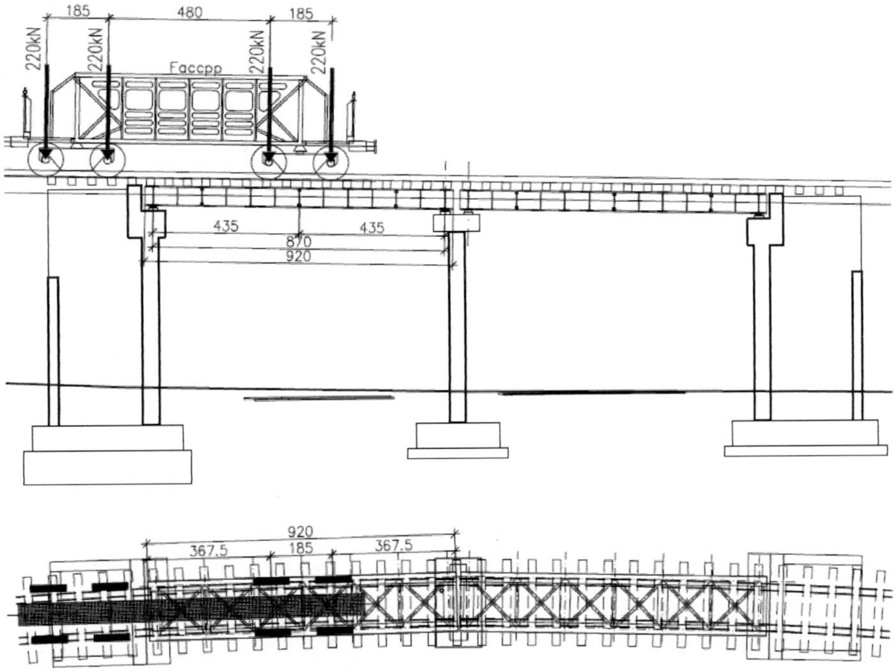

Figure 3. Load case 1

3.4. Measured parameters

The superstructure of the bridge consists of hot-rolled steel sections. For this reason, it is considered that the measurement of the deflections will give a sufficiently reliable picture of the behaviour of the structure under the test static load and it is not necessary the relative strains in the main beams to be measured. However, electrical resistance strain gauges along the lower flange of one of the main beams were installed. This was done in the middle of the span, where the maximum values of the bending moments were expected.

3.5. Testing process

In [5] it is regulated that each load case should be carried out twice, and the differences in the results should be within 10%. For the bridge case study, due to the two identical spans of the structures, and the necessity for traffic restrictions during testing, it was decided that each of the two load cases would be conducted once.

4. Results and discussion

As a result of the analysis of the measured deflection of the beam 1, loose contact between parts of the bearings and the concrete cap beam was detected. The gap was about 4.4 mm. After recalculation of the values while taking into account the additional displacement in the area of the bearing support, relatively adequate compliance with the theoretical values was evaluated.

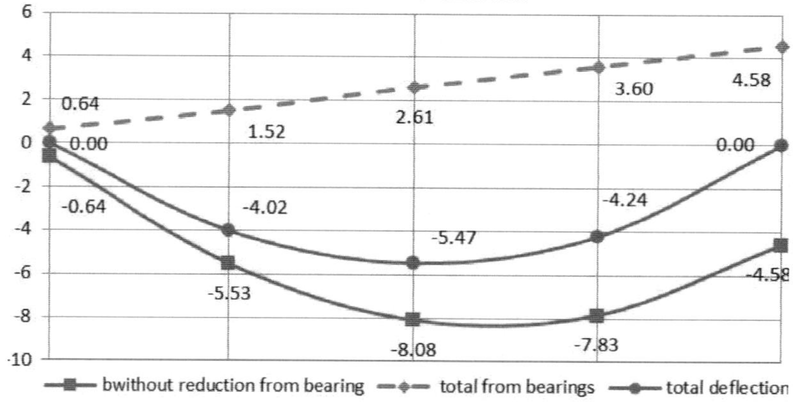

Figure 4. Deflection of the beam 1

For the other span, the measured deflections show identical values with the theoretical model.

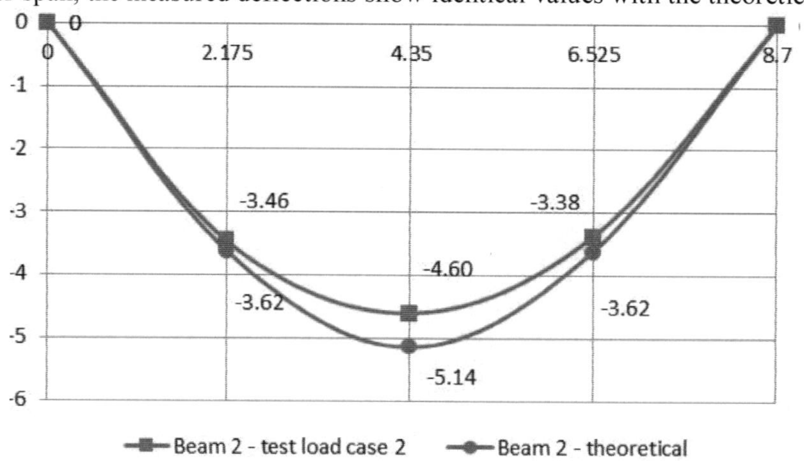

Figure 5. Deflection of the beam 2

The following conclusions can be drawn for the temporary railway bridge from the obtained in-situ test values as compared to the theoretical model:

- the bridge structure shows a good load behaviour;
- the measured displacements are smaller than the respective theoretical values;
- the differences are relatively small, which was expected, considering that the structure is entirely made of steel and is in an elastic state;
- the obtained values of the parameters are in accordance with the respective theoretical values.

5. Conclusions

On the basis of the results of the in-situ tests for the studied temporary bridge, the author recommends that such structures (with spans smaller then 15m) also need to be tested before being put into operation. Thus, problems or inaccuracies due to construction can be detected.

Even though temporary bridges typically have shorter spans, they have relatively low self-weight and the main strains are caused by live loads. Therefore, bridge testing is particularly important to ensure the proper performance of such structures. Hereby it can be advisable, for the design of relatively small self-weight structures, additional load case representing the possibility of support on a less number of supports to be taken in account.

References

[1] Nikolov P *Replacement of an old steel railway overpass without traffic interruption, Proceedings of the 7th International Conference on Road and Rail Infrastructures* – CETRA 2022, 11–13 May 2022, Pula, Croatia, ISSN 1848-9850, pp.531-537

[2] Nikolov P *Dismantling an old steel railway bridge near the village of Kalotina*, Annual of the University of Architecture, Civil engineering and Geodesy, Sofia, Volume 56, Issue 1, 2023, ISSN 2534-9759, pp.83-88

[3] *BDS EN 1991 – 2 : Traffic loads on bridges*

[4] Uzunov A 2008 *Universal temporary bridging with upper laying for spans up to 10m and velosity 50-60 km/h, Sofia*

[5] NKJI 2009 *Methodology for field testing of railway bridges*

Technological methods for pavement block repairs in road bridges

Evgenia Bogdanova

Chief Assist. Prof. Dr. Eng., Dept. "Construction Technology and Mechanisation", UACEG, 1 H. Smirnenski Blvd., Sofia 1046

e-mail address: evg_bogdanova@abv.bg

Abstract. Pavement block is a part of bridge a bridge superstructure. During the service life damage accumulates and there is need to repair. The aim of the article is to discuss the types of pavement blocks and to present the results of a study on the most common design solutions for realisation of the repair works. Factors influencing the choice of construction technology are systematized. Information about novelty formwork solutions and types of equipment for working at heights in Bulgaria is provided. A particular example of a bridge structure is presented and a conclusion is drawn regarding the preferred method of work in our country.

1. Introduction

The pavement block is the essential part of the bridge superstructure. During the operation of the facility, it is damaged by the aggressive additives, used at roads' maintenance in winter conditions, as well as by the poor or non-functioning drainage of the facility system, caused by the aging of the materials or by the appearance of cracks, caused by the dynamic loadings, etc. In order to fulfill their functions, the blocks need to be repaired. In some cases, the repair of pavement block is a part of bridge rehabilitation in order to fulfil the requirements of present design standards [8].

Generally, the repair consists of removing the damaged concrete, installing new reinforcing and re-pouring new concrete to complete the new paving block.

2. Types of paving blocks in regard to the implementation technology

The variant solutions of the pavement blocks are: monolithic concrete pavement (A), pavement with an assembly facade cornice (B) or with an assembly L-shaped cornice (C) [1]. See. fig. 1.

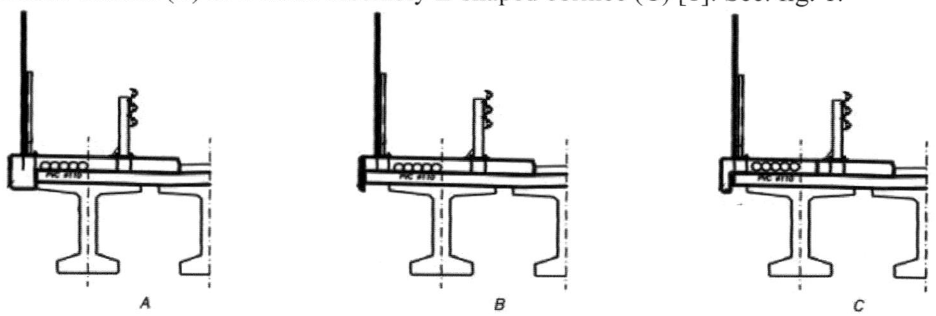

Figure 1. Paving blocks types

According to a study, provided by the author, in the public procurement for carrying out basic repairs and reconstructions of bridge facilities on highways, classified as I and II class roads, are planned to implement sidewalks, mainly with an assembly facade cornice (Type B) see fig. 2. For the other classes of roads, the monolithic method (Type A) is used for the pavement blocks. L-shaped mounting bracket cornice (Type C) is not being used.

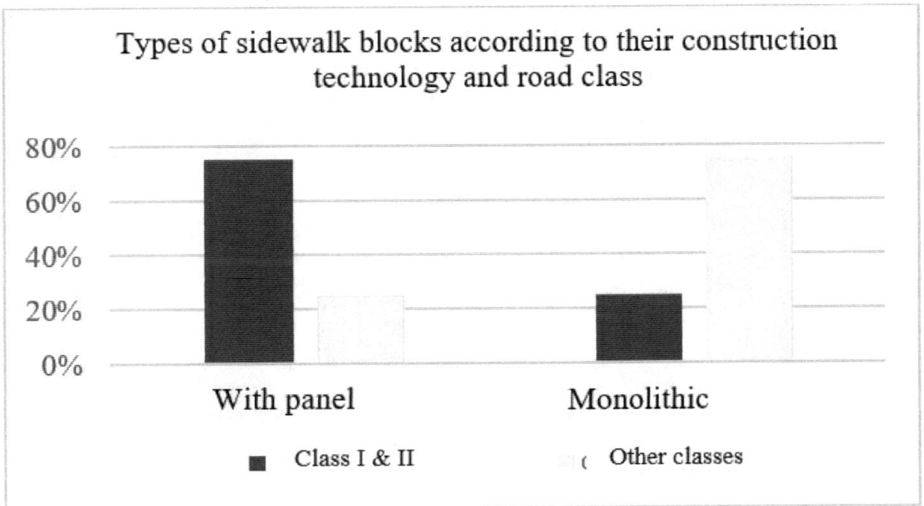

Figure 2. Design solutions for types of pavement blocks, used in the repair works at the second quarter of 2022 in Bulgaria

When executed with a facade mounting cornice, variants of polymer concrete facade panels (D) and concrete ones (E) are provided. See. fig. 3. The functions of the two panels are identical. The difference is in the element's own weight, and according to the manufacturer's technical data, the weight of the unreinforced polymer concrete is 2100 kg/cubic meter, and this of the unreinforced concrete is 2400 kg/cubic meter. In 2022 in Bulgaria there is only one manufacturer of polymer concrete panels meeting the regulatory requirements.

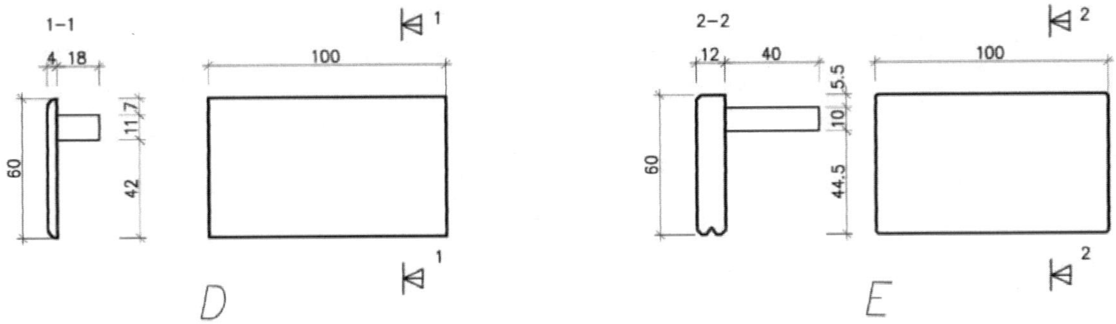

Figure 3. Facade assembly cornice

3. Factors influencing the implementation technology and system solutions

The engineering decisions regarding the technology of implementation, that must be set, depend on the following aspects:

- Geometric characteristics of the pavement block. – Factor 1 (F1)

- The geometric characteristics of the paving block refer to the cornice height and the cantilever section that needs to be completed. These characteristics determine the loads of the formwork, its sizing and implementation technology.
- Working height (the level at which the construction works are performed). – Factor 2 (F2)
- The working height is determined by the location of the element. In all cases, the beginning of the pavement block is at ground level, as it changes along the length of the facility.
- The terrain accessibility (availability of a road, water obstacle, etc.). – Factor 3 (F3)
- Accessibility to the pavement block depends on whether the bridges crosses water obstacle, a road, railway, valley, etc. And it depends on the possibility to deploy construction machinery and equipment in this section.
- Requirement to partially/or completely interrupts the traffic during the repair works. – Factor 4 (F4)
- The traffic is not allowed to be stopped during some repair works. This limitation leads to a need for precise choice of the implementation technology.
- Dimensions of the bridge. – Factor 5 (F5)
- The size of the bridge is determined by the gauge of the roadway, number of lanes, presence of dividing lines. /strips/
- Availability of power supply through an air duct [2]. – Factor 6 (F6)
- The availability of electricity is an additional limiting condition for the operation. The implementation technology has to comply with the minimal distance for safe work.

Each factor needs to be specified.

4. Some Formwork solutions and types of equipment for working at heights in Bulgaria

4.1. Types of system formwork for execution of the cantilever part of a pavement block

Companies producing formwork equipment have developed systems serving the monolithic construction of the pavement block. The principle was followed that approximately 70% basic elements, 20% system elements and 10% non-standard elements should be used for the formwork of the block [3]. The system formwork used in the execution of the cantilever and cornice part of the pavement block is a formwork carriage and system consoles fig. 4. The system consoles differs by the means they are fastened, the mechanism of receiving and distributing the loads.

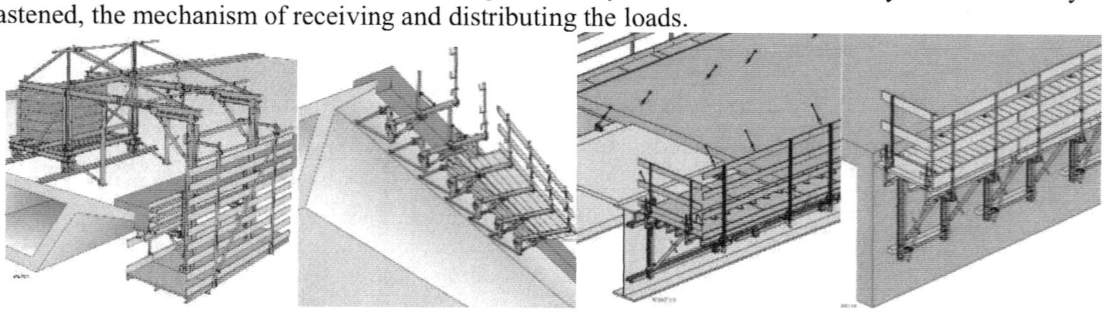

1 2 3 4

Figure 4. System formwork [4]

The possibilities of using a multi-station (linear) platform assembled from system components, see fig. 4.3, is a novelty for Bulgaria. The system allows works at the level of the construction, with the advantage of being applied both to facilities with reinforced concrete beams and to those with steel beams. The system is adapted for easy combination with beam formwork, which optimizes costs. The formwork modules can be prepared in production workshops, which shorten the period and the influence of climatic conditions. In addition, to system formwork solutions, non-system solutions can also be applied, as well as a combination of the two options, as it is necessary to prepare an individual project. See. fig. 5.

Figure 5. Individual formwork

4.2. Work at height

When carrying out repair works on bridges, very often, in addition to the pavement block, others bearing elements – such as slabs, beams, pillars, bearings - are also included. For this reason, it is necessary to make a complex decision when working at height.

The work can be single-station or multi-station (linear), regardless of the location of the equipment and works' facilities.

4.2.1 At the ground level

- The work at the ground level is carried out when there is accessibility, certain flatness and when relevant measures under OSH are ensured.
- Work from a scaffold positioned on the ground.
- Work from a cherry pickers positioned on the ground.

Different types of auto cherry pickers are available in Bulgaria. Most often, the truck lifts are encountered in the repair works. They are vehicle with a built-in lifting mechanism, having a basket or a mobile platform, used to provide temporary access to people or equipment over large horizontal and vertical distances.

An alternative option is the use of self-propelled cherry pickers. The drive of self-propelled cherry pickers could be electric, with fuel engines and/or hybrid. An additional advantage is that a licensed operator of the equipment is not required. The operation is performed after training of the workers. According to a study in Bulgaria, provided by the author, the self-propelled cranes with the greatest range of action are the articulated cranes Genie S125 and Genie Z135.

4.2.2. Different variants for working at the construction level are presented on fig. 6.

- Working with a crane equipped with a man basket.
- Work on a tower cherry pickers positioned on the structure.
- Operation from a vehicle with an access platform.
- Work platforms at the construction level.
- Work scaffolds at the construction level.

In our country, very often, for the repair works of the bridge equipment and in works at the level of the structure a crane is used, equipped with a basket to carry people, which works from a car with an access platform, [7].

Figure 6. Single-station and multi-station work equipment [5], [6]

4.3. Methods for laying the concrete mixture

- Concrete pump from structure level or ground level
- Directly from the vehicle
- Mobile crane with Crane Concrete Bucket at the level of the terrain or at the level of the structure
- Using a stationary concrete pump

During repair activities related to the implementation of a new pavement block of a bridge superstructure, depending on the season, it is possible to work in an unfavorable temperature-humidity regime. When widening of an existing road bridges, in some cases it is necessary to use corresponding larger formwork forms, and then it is of interest to analyze the influence of temperature using digital methods to determine the horizontal pressure of freshly poured concrete mix.

5. Case Study of a Reinforced-concrete Bridge

The decision under consideration is for a road bridge over the Iskar River. During operation, the pavement block is severely damaged as a result of poor drainage of the facility and caused by the aging of the materials. According to the project, a monolithic implementation of the pavement block was designed.

5.1. General features for the project:

- Located above a river (F3) fig. 7.1;
- Nearby there is an electrical supply via an air duct - a distance of about 2 m parallel along the entire length of the facility (F6) fig. 7.2;
- Small gauge of the roadway - 4m (F5) fig. 7.3;
- Stopping traffic during the stage of repair of the layers of the road surface (F4)
- Limited budget

1 2 3

Figure 7. Placement and characteristic features of a bridge over the river Iskar, Bulgaria

5.2. Decision to implement the pavement block of the bridge over the Iskar River:

After analyzing the characteristic features and studying the economically most expedient options, the decision to carry out repair works on the pavement block included the work from the ground level, which is realized with the help of a GenieS85 self-propelled crane and the use of individual non-system formwork is made (see fig. 8).
The technological method for the implementation of concrete works is performed with the help of a concrete mixer, and the supply of concrete is directly from the mixer.

Figure 8. Technological solution for the implementation of a pavement block

6. Conclusion

During the repair activities of each bridge, it is necessary to make a number of decisions concerning the implementation technology. Practice shows that, in most cases, in the technical projects there is a lack of specifics about the technology, and this task is assigned to the relevant technical departments. Also, many of the construction organizations do not trust the innovative technological systems and prefer to work with more conservative methods.
Each project is unique, with its individual characteristics, and in some cases it is not possible to give different variants for the engineering solution, caused by its specific features.
In search of the best execution technology for the formwork and concreting of a pavement block of a bridge, it is necessary, even at the level of the design, to conduct a deep analysis of the methods of execution, which would lead to the most economically expedient choice, to healthy and safe working conditions. In this way, the emphasis will be placed on the complex solution for the implementation of repair works of a given bridge.

Acknowledgments
The author would like to express his acknowledgements to the Research, Consultancy and Design Centre (RCDC) of the UACEG, Sofia, Bulgaria for financing the contract BN-288-23.

References

[1] Ivanchev J, Topurov K, 2002 *Reinforced concrete bridges*, ABC Technique
[2] Works Road infrastructure agency 2014 *Technical specification*, Ministry of Regional Development and Public
[3] PERI 2010 Variokit
[4] DOKA: 01/2021 Forming wagon T, Bridge formwork ParaTop 08/2020, Platforms assembled from system components 05/2022, Technical Newsletter – 854, 01/2016, Bridge edge beam formwork T
[5] www.moog-online.com, 5.11.2022 https://www.moog-online.com/us/products/mbl-1200-t/, https://www.moog-online.com/en/products/mfg-40-1/,https://www.moog-online.com/en/products/mbi-180-17-s-truck-mounted/

[6] https://www.asb-portal.cz/stavebnictvi/inzenyrske-stavby/mosty/dauha-napjaty-termin-a-vystavba-za-bezneho-provozu/attachment/06-al-bustan-street-south-doha-06-peri-300cmyk
[7] Nikolov P, 2019 *Some features of geometric design of bridges*, Annual of UACEG vol 52 issue S1 p 1369
[8] Jiponov A., Georgiev L., "*Reinforced structures type "filler Beam" - Advantages in reconstruction of bridge superstructures*", 2018 IOP Conference Series: Materials Science and Engineering; 9th International Symposium on Steel Bridges, 10-11 September 2018, Prague; 2018 IOP Conf. Ser.: Mater. Sci. Eng. 419 012002, https://doi.org/10.1088/1757-899X/419/1/012002, DOI:10.1088/1757-899X/419/1/012002

IOP Conf. Series: Materials Science and Engineering 1297 (2023) 012005 doi:10.1088/1757-899X/1297/1/012005

About effect of dynamic loading caused by concrete pump on bending moments, anchor forces and deformations in single level anchored retaining wall

Iliyan Markov[1], Kalin Radlov[2]

[1]Chief Assistant Prof. Dr. Eng., Department of Geotechnics, University of Architecture,Civil Engineering and Geodesy (UACEG), 1 Hristo Simirnenski Blvd., Sofia, Bulgaria
[2]Assoc. Prof. Dr. Eng, Department of Construction Technology and Mechanisation, University of Architecture, Civil Engineering and Geodesy (UACEG), 1 Hristo Simirnenski Blvd., Sofia, Bulgaria

imarkov_fte@uacg.bg

Abstract. This paper explore effect of dynamic loading caused by concrete pump on bending moments, anchor forces and deformations in single level anchored retaining wall. Dynamic loading is obtained by using a special developed dynamic mathematical model of truck-mounted concrete pump, by means of which are calculated the maximal dynamic loadings (acceleration and forces), which are transferred from concrete pump to the soil, and from there accordingly – from soil to the retaining wall under consideration. After that is performed a detailed finite element analysis of the retaining wall, by using software product Plaxis 3D, during which are analyzed the variation of bending moments, anchor forces and deformations in single level anchored retaining wall and after dynamic loading applying. There is also a numerical example considered, for which are calculated the maximal dynamic loadings, which are transferred from concrete pump to the soil, and from ground supports to the retaining wall. An amplification factor (dynamic factor) due to dynamic effects is calculated, by which is necessary to be multiplied the static loadings caused by concrete pump on soil. In the conclusion is made a qualitative assessment of the dynamic influence of concrete pump on the soil and the retaining wall section and ground anchor.

1.Introduction

Concrete pumps are construction machinery, which is used for mechanized laying of concrete mixtures at building sites [1]. Truck-mounted concrete pumps represent a special type of concrete pumps (a combination between concrete pump and standard truck chassis), which are very often used during construction process of ordinary residential buildings, industrial facilities and infrastructure projects (bridges, roadways, etc.). Most of the scientific researches in the area are focused only on the dynamic loadings, which are transferred from the pulsating concrete mixture in the flexible tube to the folding boom for concrete laying, as for example: [2], [3] and [4].

On the basis of obtained vibration frequency and deformations in the folding boom of concrete pump model JJ315/45-K in [4], is possible to be calculated the approximate value of acceleration in the center of gravity of the folding boom Sa [m/s^2] (it is assumed, that center of gravity of boom is located near to the middle length of boom), by using the equation:

IOP Conf. Series: Materials Science and Engineering 1297 (2023) 012005 doi:10.1088/1757-899X/1297/1/012005

$$Sa=(2\pi*f_{JIB})^2*D/2,[m/s^2] \tag{1}$$

where $D,[m]$ is the maximal deformation in the top of the boom; $f_{JIB},[Hz]$ – fundamental natural frequency of the boom in vertical direction. It is assumed that the fundamental natural frequency of the boom in vertical direction (axis Z), in case of maximal horizontal reach of folding boom (maximal horizontal position of boom) is equal to: $f_{JIB}=0{,}54[Hz]$, and as a result of that are calculated the approximate values for accelerations and deformations in the vibrating boom of the concrete pump, which are shown in table 1 below:

Table 1. Approximate values for maximal vertical deformations (axis Z) in the top of folding boom of concrete pump model JJ315/45-K [4] and calculated maximal vibration accelerations in the center of gravity of boom, in case of pumped concrete pressure 10MPa and rotational speed of driving internal-combustion diesel engine about 1500rev/min.

	Frequency of concrete pumping (frequency of concrete pulsations in the flexible tube), [Hz]						
	0.117Hz	0.200Hz	0.233Hz	0.233Hz	0.267Hz	0.300Hz	0.383Hz
Maximal vertical deformations in the top of folding boom	0,052m	0,320m	0,350m	0,320m	0,280m	0,360m	0,320m
Maximal vibration accelerations in the center of gravity of boom	0,3 m/s²	1,84 m/s²	2,01 m/s²	1,84 m/s²	1,73 m/s²	2,07 m/s²	1,84 m/s²

It is a fact, that all above mentioned scientific researches are focused only on the dynamic loadings, which are transferred from the pulsating concrete mixture in the flexible tube to the folding boom for concrete laying, as well as monitoring and control (active control and passive control) of the dynamic vibrational behavior of folding boom. Nowhere is analyzed the case, when the concrete pump support structure is located near to fortification wall of construction excavation, by which the vibrational loadings from concrete pump ground supports (outriggers) are transferred to the fortification wall of construction excavation.

The present development is focused to survey the scenario of combined vibrational loadings on the pump ground supports (caused by unbalanced rotating parts from machine driving system and by vibrational motion of folding boom due to pulsating concrete mixture inside), which loading is transferred accordingly to the fortification wall of nearby located construction excavation. The usual engineering practice in this case is to assume an approximate value of amplification factor (dynamic factor) by which to be multiplied the static loadings from the concrete pump, which means "static approach" for task solution. The main purpose of the present development is to explore the real dynamic loadings and real dynamic effects on fortification wall, by means of dynamic analysis, as well as to calculate the real value of amplification factor (dynamic factor), by which have to be multiplied the static loadings from the concrete pump. The present development is performed according to the setting, which is presented in figure 1.

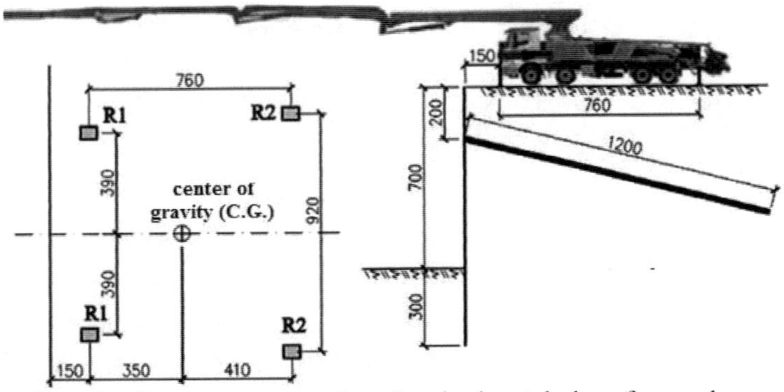

Figure 1. Overall setting by concrete pump location: horizontal plan of ground supports and cross section view

2.Development of mathematical dynamic model for deriving the differential equations of motion
It is assumed that the horizontal vibrational motion of the concrete pump ground support structure (outriggers) are negligible, because of which they are ignored for the purposes of the present development, i.e. it is considered only vertical vibrations (loadings) transfer from pump ground supports to the landfill below and nearby fortification wall. mathematical dynamic model. For the purposes of the present research, it is developed a mathematical dynamic model, which is shown on figure 2 below.

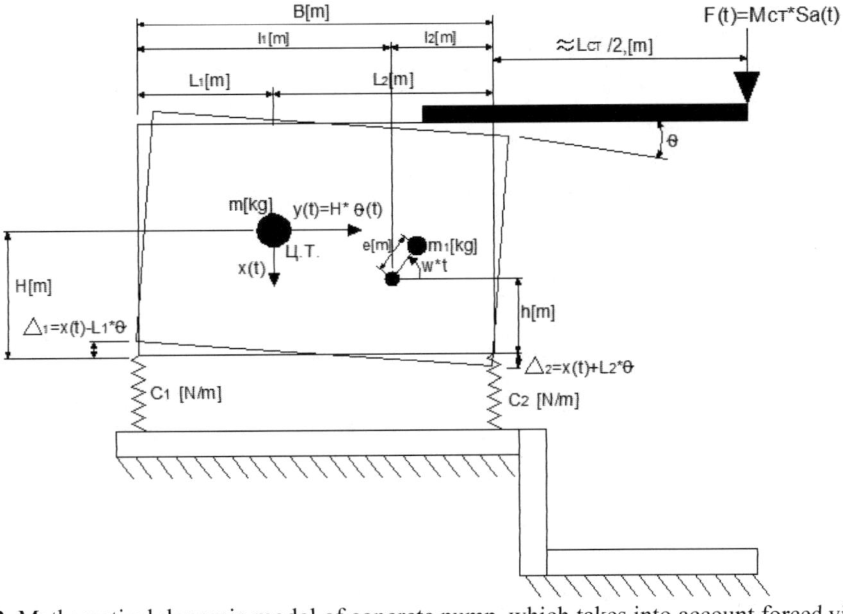

Figure 2. Mathematical dynamic model of concrete pump, which takes into account forced vibrations from two excitation sources

For the present development are used the following denotations: m[kg] – own mass of truck-mounted concrete pump; $C_1 = [N/m]$ и $C_2 = [N/m]$ – stiffness factors for the support structure in the front and rear side of truck pump; $L_1 = [m]$ and $L_2 = [m]$ – distances between center of gravity (C.G.) of truck pump and ground supports location; $l_1 = [m]$ и $l_2 = [m]$ – distances between unbalanced rotating part of pump driving and ground supports location; J [kg*m^2] – mass moment of inertia by inclination (rotation) of pump truck chassis regarding to center of gravity (C.G.) of truck pump; $\theta(t)$ – angle of inclination (rotation) of pump truck chassis; $x(t)$ – vertical translational motion of center of gravity of the truck pump; $\Delta_1[m]$ and $\Delta_2[m]$ – relative motion (deformation) of rear and front ground support of the truck pump (see figure 1); e [m] –eccentricity of unbalanced rotating part in the pump driving; h[m] – vertical distance between unbalanced rotating part and ground support surface; $\omega[rad/\sec]$ – angular velocity of unbalanced rotating part in the pump driving; $m_1[kg]$ –own mas of unbalanced rotating part in the pump driving; F [N]- dynamic force, which is caused by folding boom vibrations; $L_{CT}[m]$- maximal horizontal length of folding boom; $M_{CT}[kg]$-own mass of folding boom; $S_{a.MAX}[m/s^2]$ – maximal value (amplitude) of the acceleration in the center og gravity of the folding boom in vertical direction (axis Z); $\omega_{CT}[rad/\sec]$ – fundamental natural frequency of the folding boom in vertical direction (axis Z): $\beta[N.s/m]$ – viscous damping coefficient of pump ground support structure.

The two differential equations of motion are derived by using the Newton's law and are obtained as follows:

$$m*\ddot{x}+2\beta*\dot{x}+(c_1+c_2)*x-\beta*(L_1-L_2)*\dot{\theta}-(c_1*L_1-c_2*L_2)*\theta= \\ =m_1*e*\omega^2*\sin(\omega t)+M_{CT}*S_{a.MAX}*\sin(\omega_{CT}.t) \qquad (2)$$

$$(J+m*H^2)*\ddot{\theta}-\beta*(L_1-L_2)*\dot{x}-(c_1*L_1-c_2*L_2)*x+\beta*(L_1+L_2)*\dot{\theta}+ \\ +(c_1*L_1^2+c_2*L_2^2)*\theta=m_1*e*\omega^2*\sin(\omega t)*(L_2-l_2)+ \\ +m_1*e*\omega^2*\cos(\omega t)*h+M_{CT}*S_{a.MAX}*\left(\frac{L_{CT}}{2}+L_2\right)*\sin(\omega_{CT}.t) \qquad (3)$$

3.Solving a specific numerical optional case with example input data

There is used the following example input data: distances between pump ground supports: A1=7,8m; A2=9,2m; B=7,6m (see figure 1); m=40 tons; $L_1=4,1[m]$; $L_2=3,5[m]$; $l_1=6,0[m]$; $l_2=1,6[m]$; LCT = 48m; $\omega_{CT}=3,3912[rad/\sec]$; $M_{CT}=4200[kg]$; h=1,7m; H=1,9m; $C_1=1000[kN/m]$; $C_2=1000[kN/m]$; $S_{a.MAX}=2,5[m/s^2]$; J =30000 [kg*m^2]; $\beta=2000[N.s/m]$; e=0,01m; $\omega=157[rad/\sec]$ (i.e. 1500 rev/min); $m_1=50kg$. After solving the differential equations of motion with the example input data by using the software program Matlab, there are obtained the following results (see figure 3 and figure 4):

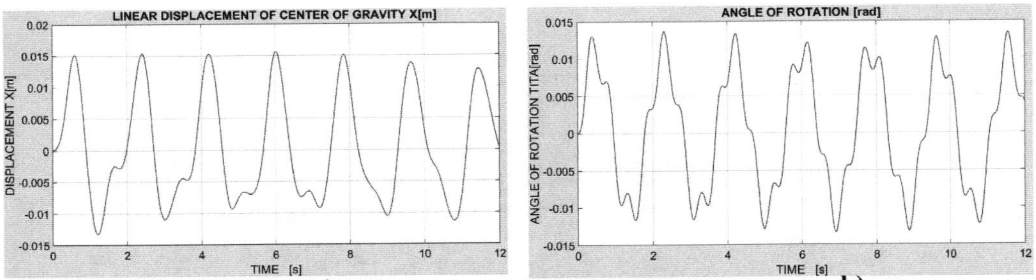

a) b)

Figure 3. Obtained results for the motion of truck pump's chassis in the first 10 seconds: a)-vertical translational motion of center of gravity of the truck pump x(t); b)-angle of rotation of truck pump's chassis $\theta(t)$

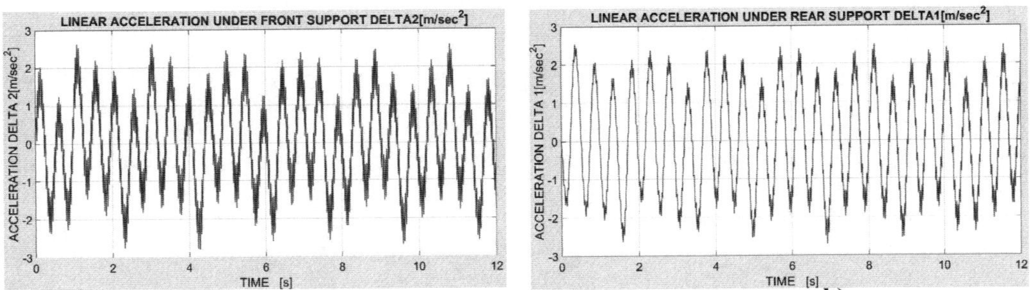

a) b)

Figure 4. Obtained results for the relative motion (deformation) of front and rear ground support of the truck pump in the first 10 seconds: a)-acceleration of front ground support of truck pump: $\ddot{\Delta}_2 = \ddot{x}(t) + L_2 * \ddot{\theta}(t)$; b)- acceleration of rear ground support of truck pump: $\ddot{\Delta}_1 = \ddot{x}(t) - L_1 * \ddot{\theta}(t)$

4. Retaining wall

In the Bulgarian building codes [5], and in Eurocode7 [6], there is no exact statement that takes into account the dynamic load, which has an impact on stresses and deformations in the soil, as well as on the structural elements. There are studies [7] that investigate the earthquake influence. A retaining wall with the cross-section shown in Fig. 1 is investigated. In order to evaluate the influence of the soil parameters, two models were investigated – one cohesive soil (clay), and on noncohesive (sand).

4.1 3D FEM model

A three-dimensional model was prepared, according to Fig. 5, in which volumetric finite elements were introduced, and the the load from the support is distributed over a square area.The boundaries are as follows: x_{min}=-30m, x_{max}=30m, y_{min}=-30m, y_{max}=30m. The minimum depth is z_{min}=-20m

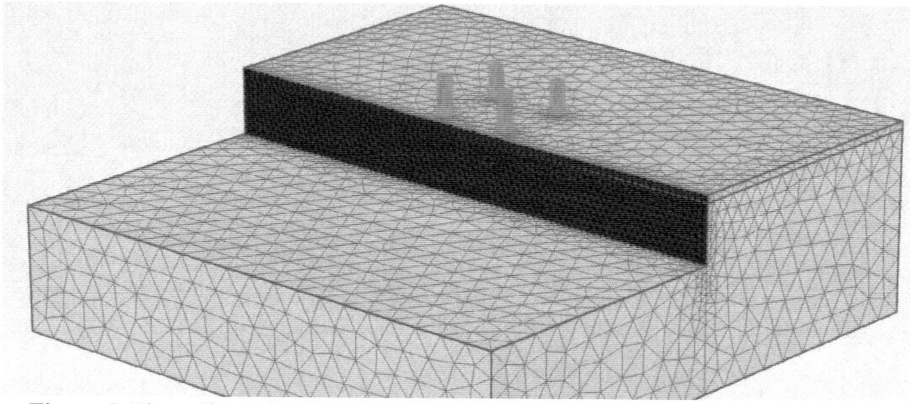

Figure 5. Three-dimensional computational model with load from the concrete pump.

Five different stages are modeled, taking into account the construction sequence: Initial phase, in which the initial stresses are generated in the soil media, Phase 1 - instalation of the wall, Phase 2 - execution of excavation up to k.-2.00 and anchor instalation, Phase 3 - execution of excavation up to k -7.00, Phase 4 – static action of the crane, Phase 5 – dynamic action of the crane. The load position is in accordance with Fig.1

4.2 Soil model

The research was done for two types of soil - sand and clay, with soil properties shown in Table. 2. The adopted soil model is HSMsmall, because it takes into account the stiffness of the soil at relatively small deformations - those expected under high-frequency dynamic loading.

Table 2. Soil characteristics, according to the selected HSMsmall model

	Volume weight	Internal friction angle	Cohesion	Secant stiffness	Tangent stiffness for primary oedometer loading	Unloading / reloading stiffness	Threshold shear strain	Reference shear modulus at very small strains
	γ [kPa]	φ [°]	c [kPa]	E_{50}^{ref} [kN/m²]	E_{oed}^{ref} [kN/m²]	E_{ur}^{ref} [kN/m²]	γ_{07} [-]	G_0^{ref} [kN/m²]
Sand	18.50	32	3	30E3	30E3	90E3	0.15E-3	120E3
Clay	16.50	18	25	20E3	20E3	50E3	0.15E-3	120E3

4.3 Defining structural elements

In Bulgaria a lot of retaining structures are executed from steel profiles and wooden sections between them, so called "berliner" wall. Therefore a wall with a section HEA240/80cm was adopted in this study. Material properties for plate element in Plaxis 3D are: thickness d=24cм, γ=2,6кN/m^3, E_1=1.3E7, E_2=1.3E5, ν_{12}=0, G_{12}=6.5E6, G_{13}=2.7E6, G_{23}=2.65E5. The ground anchor is modeled using embedded beam with material properties E=31,3E6 кN/m2, γ=25 кN/m^3. Diameter of grout body is D=0.138cm.

4.4 Loading conditions

The concrete pump exerts a load which is assumed according to Fig.1 where R1=81kN, R2=69kN. The dynamic component is obtained by multiplying the static vertical load by a time-varying dynamic

IOP Conf. Series: Materials Science and Engineering 1297 (2023) 012005 doi:10.1088/1757-899X/1297/1/012005

multiplier. The dynamic multiplier is calculated according to the accelerations shown in Fig.4

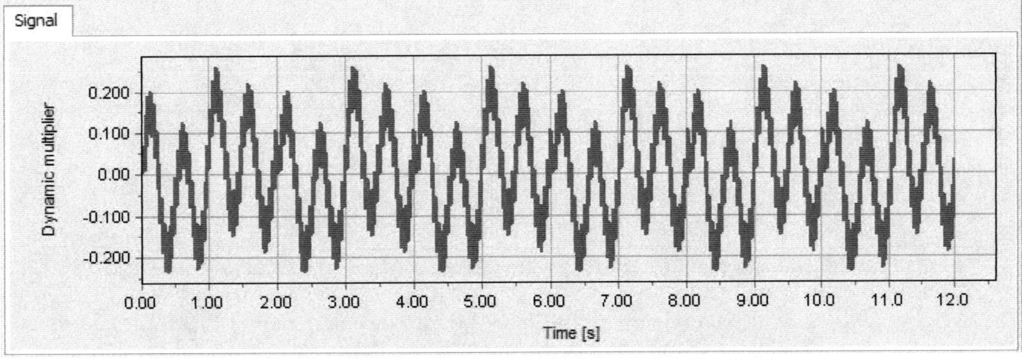

Figure 6. Dynamic multiplier for front support for 12s dynamic time

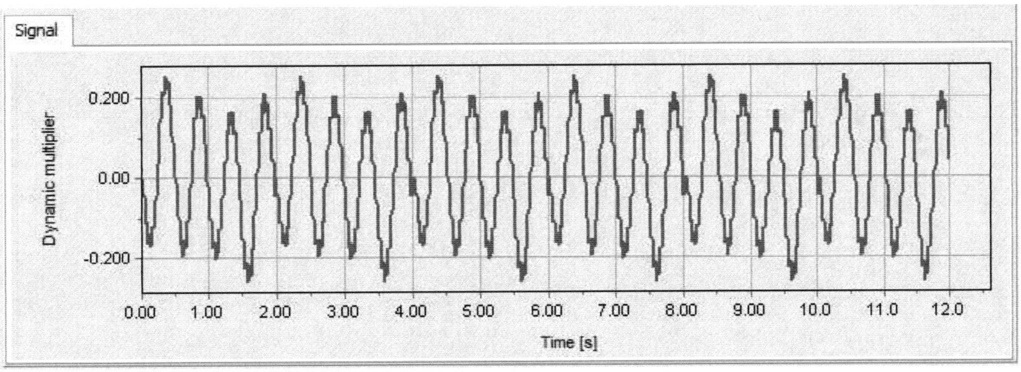

Figure 7. Dynamic multiplier for rear support for 12s dynamic time

From the dynamic multiplier functions, it can be noticed that the force in the front support oscillates by about ± 0.25 of 81kN, or the load amplitude is (61-102)kN, which are distributed over an area of one m2. For the rear support, the load range is (46-76) kN, again distributed over an area of one m^2. The setup of dynamic phase in PLAXIS, here can be point of interest. In model conditions the Dynamic is cheked and the boundaries are as follows: X_{min}, X_{max}, Y_{min}, Y_{max}, Z_{min} are viscous, Z_{max} is none. Dynamic time is set to 12 seconds, and Max number of step to store is 100. Numeber of substeps is 23.

5. Results from the retaining wall analisys

The purpose of this research is to make a qualitative assessment of the dynamic influence of the applied load from a concrete pump on the behavior of the soil and individual structural elements. For this reason, it was chosen to monitor the maximum bending moment of the retaining wall, the horizontal displacement, as well as the force in the anchor.

5.1 Influence of dynamic load on displacement

It was chosen to track the displacement of a center point of the soil body with coordinates (0,0,0), which is on the top on the wall on the centerline of concrete pump. The displacement in the begining of the phase is 0.8mm for sands, and 0.001mm for clays.

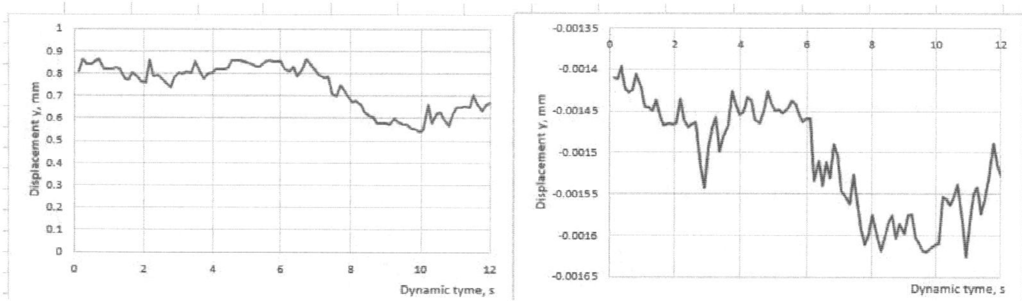

Figure 8. Displacement of central point with coordinates (0,0,0) a)Sand b) Clay

The dynamic phase provides an opportunity to track the displacement during these 12 seconds and evaluate its change depending on the load oscillation. Fig.8 shows the change of displacement of point and the range of displacement is 0.2mm for sand. For clays dynamic loading almost has no influence.

5.2 Influence of dynamic load on the anchor force.

Two anchors are monitored – one is in between pump support, exact in center line. The second is just under the pump support.

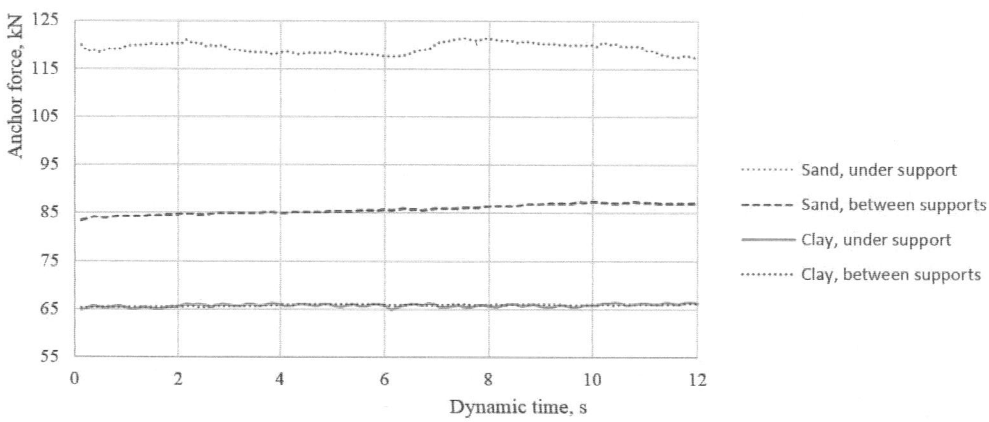

Figure 9. Change of two anchor force for clay and sand, 12s dynamic time

The change of anchor force is shown on the Fig. 9, where can be seen the change of anchor force for 2 anchor for clay and sand. Analyzing the result can be noticed, that for all the anchor there is a slightly noticeable change, but none from a practical point of view, or dynamic oscillation does not change the anchor force. The other interesting result is noticeable – for clays under support and in between support the anchor force is the same. In sand, however, the force under the support is greater than that between the supports by about 50%

5.3 Influence of dynamic load on the bending moment

The change of bending moment M_{11} in the retaining wall is traced at two points. One is at depth of 2m where is the maximum positive moment M_{max}, The second one at a depth of -5m where is the negative moment M_{min}.

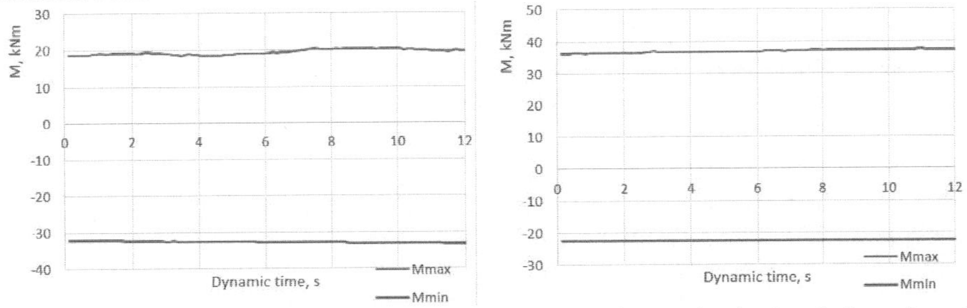

Figure 10. Change of bending moment M_{11} for points at depth -2 and -5m. a)sand b) clay

On Fig.10 is plotted change of two points for sand and clay. There is a slight change of the order of several percent, which has no practical value.

5.4 Sufficiency of dynamic signal length

A logical question would be if the accepted length of the signal is sufficient to evaluate the multi-hour action of the dynamic load of a concrete pump. Here we can set a criteria: establishing a constant value of the anchor force, bending moment and displacement of retaining wall structure during the time of action of the dynamic load; Fig.8, Fig. 8 and Fig.10 shows convergence of values of displacement .

6. Conclusions

The presented work provides a methodology for evaluating the dynamic load of a concrete pump on the stressed and deformed state of a soil, as well as on the stresses and deformations of a retaining wall and ground anchor. After a series of calculations, the results can be summarized in several points:

- The dynamic component of static force of a standard concrete pump acting for a certain time was calculated and it is in range is (61-102)kN for front support and (46-76) kN for rear support;
- Increase in force in the anchor due to the dynamic effect is negligibly small;
- Increase in displacement due to the dynamic effect is negligibly small;
- Increase in maximum and minimum bending moment due to the dynamic effect is negligibly small;

References

[1] Katov, P., Construction Machinary, Technica, 1988.
[2] Dirk Nissing, Werner Bernzen, Helmut Schwarz, On vibration control of a concrete pump, European Control Conference, Karlsruhe, Germany, August 1999, DOI: 10.23919/ECC.1999.7100046
[3] Rongsheng Liu, Yingjie Gao, Sen Yang, Yulin Yang, Vibration control of the boom system of truck-mounted concrete pump based on constant-position commandless input shaping technique, Hindawi Publishing Corporation, Shock and Vibration, Volume 2015, Article ID 420935, 9 pages, http://dx.doi.org/10.1155.2015/420935
[4] Ren, Y.; Yang, W.; Sun, X.; Zhi, J.; Li, J.; Wang, H. Vibration characteristics of concrete pump trucks with multiple postures and multiple conditions based on the secondary development of HyperWorks, Processes 2023, 11, 1483, https://doi.org/10.3390/pr11051483
[5] Bulgarian code for spread footing design
[6] Eurocode 7
[7] Mandal Tirupan, Jadhav Rahul, Deshmukh Vivek Balwantrao, Behaviour of Retaining Wall under static and dynamic active earth pressure, International Journal of Earth Sciences and Engineering

Structures for permanent stabilization of the sea coast

Professor Dr. Eng. Nikolay Zhechev
"St. Ivan Rilski" University, Department of Underground Construction
Sofia, Bulgaria.
E-mail: n_rafailov@abv.bg

Abstract: The coastal protection of the country, built in the early 1970s, has been devoid of even preventive measures such as depositing crushed rock along the coastal shoreline for several decades. Instead of prevention, an entirely ineffective defensive concept has been implemented, with concrete protective barriers accumulating along the shores and temporary breakwaters and groynes being created using additional concrete blocks. These measures, which essentially destroy the coast, are complemented by periodic sand embankments during winter, excavated from the coastal strip and piled in front of iconic coastal establishments. All these activities lead to terrain destabilization and erosion of the country's coastline and sandy shore.
The planned embankment along the Black Sea coast is not a solution to the problem either. Firstly, this dam is insufficient for regions like Balchik to create the necessary counterbalance to ensure the stability of the landslide slope. Secondly, the practice where such structures have been built shows that landslides continue due to ongoing coastal erosion caused by the energy of waves passing through the dam. Thirdly, this type of dam makes the beaches difficult to access and unattractive due to the lack of sand.
After analyzing the current state and the reasons for ongoing processes of slope destabilization and the loss of sandy beaches along the Black Sea coast, measures for sustainable coastal protection are proposed. Conclusions and recommendations for practice have been drawn.

1. Introduction

The initial coastal protection measures, built in the early 1970s and currently at some critical locations, are being supplemented with large rock fragments and concrete structures. The constructed rock and concrete breakwater dissipate wave energy but allow water to pass through, causing additional erosion of the adjacent slope (Figure 1).

When the lack of any other measures, such as periodic replenishment of the coastal shoreline with sediments and the neglect of existing, unmaintained protective structures like groynes, sea walls, and the activation of landslides due to groundwater and undermining caused by urban development (Figure 2), are added to the problem, coastal reinforcement becomes a struggle to preserve investments in buildings along the coastal strip.

Figure 1. The coastal street of Balchik town. The street is sinking, and the additional rock filling now exceeds the height of the protective parapet, which has become unnecessary and dismantled. A palliative attempt to stabilize the sinking groyne by adding concrete blocks - disappearing groynes are a widespread case along the Bulgarian coastline.

Figure 2. The development of infrastructure causes the undermining of slopes, activating the movement of soil masses.

The continuous action of waves causes the accumulation of sediment and landslide material on the fifth of the slope. They are susceptible to coastal erosion, leading to further destabilization of the coastal slope, loss of land, cracking of buildings, and the disappearance of the sandy shoreline.

The proposal of a project by "Geozashtita - Varna" for a breakwater-dam along the coastline to stabilize the coastal area is fundamentally a correct decision, but there is a risk that the necessary counterbalance from retaining forces may not be achieved, and the observed negative phenomena may persist (Mocket I., Simm J., 2018). The documented cases of adding rocky material have been unsuccessful. The assault on the coastal strip by private entrepreneurs is unregulated and solely driven by the principle of getting as close as possible to the coastal level. This leads to the subjugation of technical criteria for stable slopes and the preservation of the beach area to entrepreneurial-architectural interests. Obviously, the management of coastal cities follows events, merely legalizing investment intentions, and cannot control the escalating processes of coastal erosion, which also have social repercussions (Zhechev N., 2021).

Figure. 3: Breakwaters are often constructed in front of seaside restaurants using excavated sand in the coastal contact zone with the sea. The result is a reduced beach area and transported sand forming shallows along the coastline.

The shape of the existing coastal protection structures is unfavorable for retaining fine-grained beach deposits. Before the implementation of the aforementioned measures, the area had a natural sandy beach, which is now largely absent except for some parts of the coast under favorable sea conditions. The depletion of surface beach deposits, exposing the main geology to marine abrasion, is also due to the reflected wave energy from the retaining wall structure, throwing away the beach deposits.

The consequences of prolonged coastal erosion in areas like Balchik are twofold. Firstly, on a local scale, erosion and instability lead to further declines in coastal slopes, putting residential properties, access ways, pedestrian paths, and services occupying the slope and intermediate zones at risk. Secondly, the consequences result in the loss of material in the lower sections of the landslide complex. Observations of the ongoing movement in the area indicate that this mechanism has already reached its critical equilibrium. Therefore, the loss of additional material would negatively impact the stability of the zone and increase the speed of movement of the landslide circus (Rafailov R., 2022).

The practice of constructing sand groynes along the beaches before winter exacerbates the destructive processes. Similar groynes are used in major European resorts without harming the beaches. However, in our country, the negative results in our country derive from the strategy of concessionaires not to protect the sand on the beach, but to protect the restaurants on the beach from the waves. Under this strategy, sand is often excavated from the area in contact with the sea (Fig. 3). In some cases, excavators even partially enter the sea. The saying of these concessionaires, "The sea takes it, but then it gives it back," is not proven, as it is uncertain whether the sand will be returned to the same beach by the waves. In most cases, the sand forms shallow areas near the shore, and after the winter months, the beach is "eaten up" for the next summer. An example of the results of this detrimental strategy is the beaches in the area of St. St. "Constantine and Elena," but the phenomenon is typical for the entire Bulgarian coastline.

The analysis leads to the conclusion that the groynes constructed along the beaches in this unregulated manner are harmful to the coastal strip and create conditions for landslide processes.

To stop the practice of building groynes in a way that benefits restaurateurs but damages the coastal strip and slope stability, strict rules for their use should be introduced. The strategy of these rules should be that sand groynes are permissible without harming the beaches and even contributing to the accumulation of additional sand. The rules should not only include construction requirements but also take into account the meteorological conditions and when the sea throws sand ashore. The delivered sand should be periodically drawn back to the rear of the beach. This sand should be piled up in areas where the sea has eroded the shore after a storm. This way, a constant increase in the security of the adjacent slopes, preservation, and expansion of the coastal strip would be ensured, which is also in the interest of the concessionaires.

2. Proposals for constructive measures for the protection of the sea line

In cases of triggered landslide events, the measures can be summarized into four main types:

1. Shoreline protection through coastal slope activities.
2. Protection using coastal defense walls.
3. Investigation and, if necessary, stabilization of all slope sections that have undergone excavation without justification.
4. Preemptive stability assessments of the area prior to future development.

This report will focus on the specifics of the first two measures, for which there are numerous precedents of unsuccessful implementation. The third and fourth types of measures are of a general geotechnical nature and require more standardized investigation approaches, which have been established in practice.

- **Shoreline protection through activities related to coastal slope management.**

The main objective is to establish effective measures for shoreline protection, aiming to prevent material erosion from the toe of the coastal slope. These measures maintain the level of the toe in the overall stabilization system and also limit the extent of local instability. Additionally, the proposed scheme stabilizes the slope, enhances safety, safeguards properties, and restores the coastal promenade, which in some areas like Balchik town is already facing irreversible loss. The overall plan of the proposed measures is illustrated in Figure 4.

IOP Conf. Series: Materials Science and Engineering 1297 (2023) 012006 doi:10.1088/1757-899X/1297/1/012006

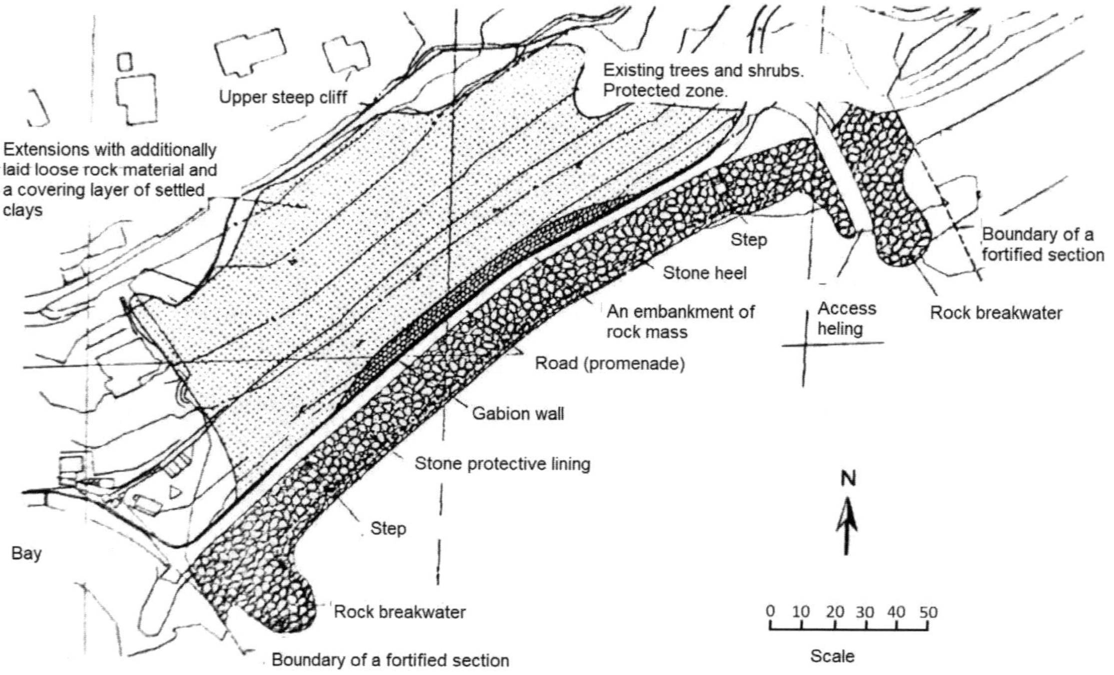

Figure 4. General plan of the proposed measures.

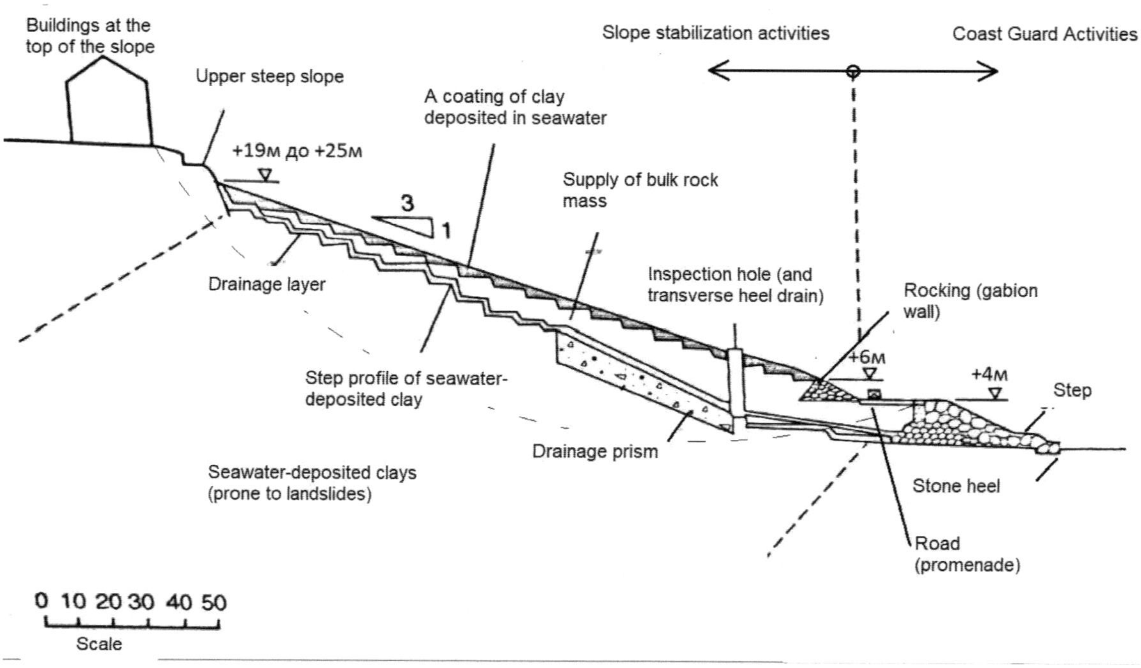

Figure 5. Characteristic section during activities on the sea slope.

Figure 5 shows a cross-sectional profile in the coastal protection zone. The scheme of the proposed measures includes the following main elements:

• Facing the toe of the slope with rock mass to protect against marine erosion and abrasion.

• Slope stabilization through regrading with additional filling and underpinning.

• Deep and shallow drainage measures.

• Accessible promenade at the back of the facing, constructed along the entire length of the structure.

• Stairs to facilitate maintenance activities on the facing and the shoreline.

The purpose of the proposed regrading, combined with drainage and compacting, is to increase the safety factor and create a shallower slip surface that terminates within the volume of the faced embankment. A common requirement is that the faced embankment and the adjacent coastal street should be at least 4m above sea level. By using the facing, seepage from the sea is halted, and the increased surface area and volume of the drains direct groundwater away from the sliding area.

However, in areas like Balchik, this condition is not met, as the height of the coastal street above the sea is around 2m. Due to the consequences of such cases, development as a location and scale can proceed only after implementing the described reinforcement and drainage activities. In the cases of Balchik and other parts of the coast, construction and excavation are leading activities, hindering the necessary stabilization measures for the slope and the adjacent infrastructure, and resulting in compromised stability and rapid deterioration. This fact confirms the assertion that building densification in the lower zone of the slopes does not always increase safety and highlights the need to consider the entire range of interventions carried out.

Shoreline protection through coastal defense walls.

The use of coastal defense walls as wave-breaking structures in current practice mainly leads to negative outcomes. For instance, in the case of Krapets, located on the northern Black Sea coastline, these results include sand erosion, destruction of beaches, slope erosion, deepening of the shoreline, and its transformation into a hazardous zone for use. In most cases worldwide, the walls are overwhelmed by sea waves, leading to further erosion of the coastlines, slope collapses, and negative impacts on marine populations. Despite the logical concept of using preventive structures to avoid these negative consequences, coastal defense walls remain a preferred technical solution for construction contractors due to their ability to accommodate increased capital investments and their implementation using relatively conventional construction methods. For this reason, walls will continue to be utilized.

In this context, the aim of this proposal is to improve coastal protection with walls and preserve the coastline. Two main cases from practice will be examined – steep and moderately sloped terrains.

The common goal between the two cases is to restore the lost volume of the coast and stabilize the adjacent slope. The restored section of the coast can serve as a coastal street or road.

In the first case, there are significant collapses due to marine abrasion of the original slope profile, and the cliff is dangerously overhanging. The coastline, as the wave zone, continues to erode, and sand is excavated due to storms (Figure 6). In such harsh conditions,

IOP Conf. Series: Materials Science and Engineering 1297 (2023) 012006 doi:10.1088/1757-899X/1297/1/012006

the only suitable solution to save the coastline is the construction of retaining walls. These walls should meet the following conditions: [Further details about the specific conditions and requirements for the retaining walls are likely to follow in the original text.]

- To ensure the overall stability of the adjacent slopes.
- To serve as protective barriers against accidental rockfall, while larger rock blocks should be reinforced with anchored nets or steel/reinforced concrete elements.
- In addition to functioning as wave-breaking structures and supporting the counterfort for overall slope stability, the supporting structures must also provide local stability for the road infrastructure, which means increased height, for example, 5.80 meters.
- They should not be conventional wave-breaking facilities that absorb the entire energy of the waves. Instead, wave dissipation should be combined with the restored beach zone to provide protection to the toe of the structure.
- The foundation of the coastal defense wall should be laid on a solid rock base or below the leveling level using pile foundations.

The shown coastal defense wall in Figure 6 can provide direct protection to a road or coastal street. A similar principle was used in the construction of the coastal street at the "Raiksa Gradina" complex in Sveti Vlas.

The actual defense of the steep rocky slope consists of a concave wall facing the sea and a low limiting/barrier wall towards the slope. The materials used for constructing the wall can be sandstone blocks with a minimum thickness of 650 mm or reinforced concrete, with backfilling of rocky material behind the wall to support the infrastructure and slope stability. The wave-breaking wall itself rests on a stable foundation. Additionally, a small beach is provided to dampen the impact of the waves and offer some protection to the toe of the structure.

When possible, effective coastal protection can be achieved through the accumulation of a sandy slope. However, in these cases, sand should not be directly piled onto the exposed foundation. In many places, concessionaires partially address this issue by adding a thin layer of sand each summer season, which the sea washes away, causing continued erosion of the coastline.

In such cases, it is recommended to create a sandy slope on a stepped, inclined wall made of monolithic and assembled elements (Figure 7). The use of protective piles is advisable for the foundation of the structure. This approach preserves beach access, prevents wave energy, accumulates sand, preserves road infrastructure, stops beach erosion due to seepage, and ensures slope stability. It is recommended to raise the beach level in front of the wall to +4.50 meters. Similar to the first type of measures, for coastal streets like Balchik, this type of coastal protection is highly suitable. It means that the street should be raised by approximately 3 meters, which is currently impossible due to the already built-up coastline. Therefore, development should follow the implementation of coastal reinforcement activities.

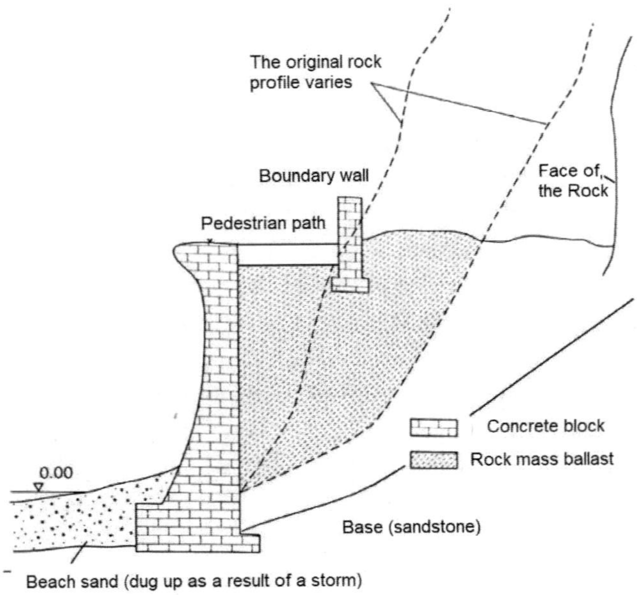

Figure 6. Typical cross-section through a coastal defense wall on a steeply sloped terrain.

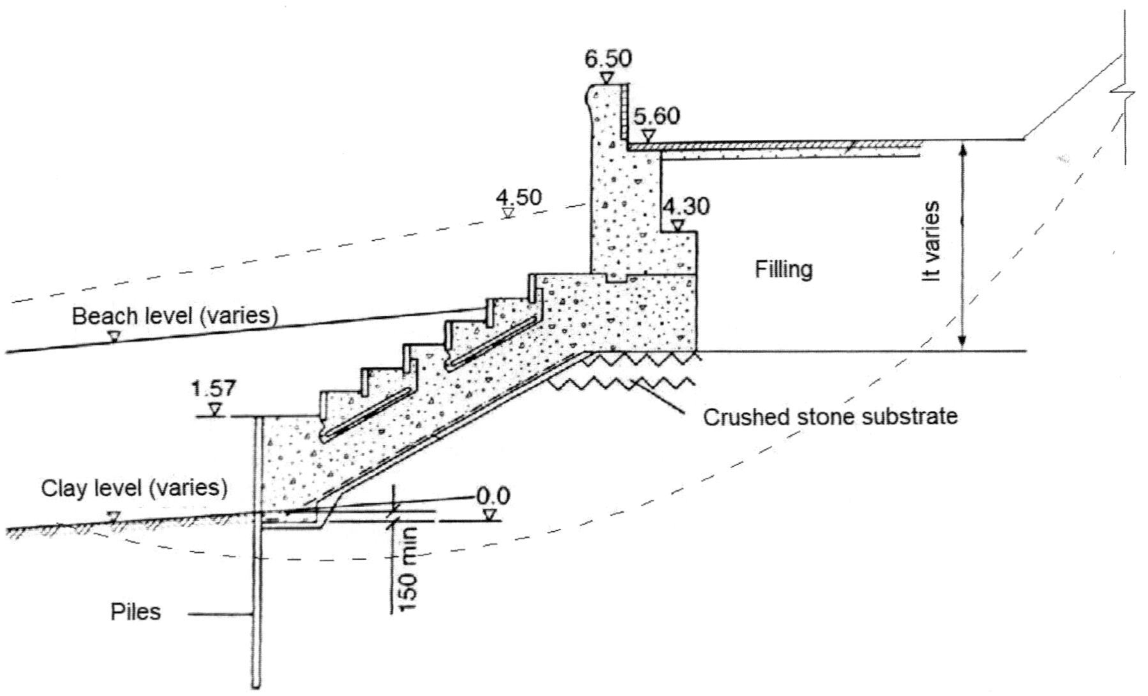

Figure 7. Detail for creating permanent coastal reinforcement with a durable sandy embankment on a moderately sloped terrain.

3. Conclusion

From the conducted analyses and proposals for coastal reinforcement, the following conclusions and recommendations can be synthesized:

3.1 An overview of current coastal defense marine structures in the country and the consequences of ineffective practices in this field has been conducted.

3.2 Guidelines for constructing sand embankments have been derived to neutralize their harmful impact on the coastline.

3.3 The specific constructive measures proposed for the slopes and through coastal defense walls provide preventive protection for the coast and road infrastructure, usability of the beach area, and stability of adjacent slopes.

3.4 The practice of pre-development of the coastal area followed by seeking ways to stabilize it should be discontinued. Inappropriate development and excavation of terrains are the main causes of lost beaches, coasts, and triggered landslide processes. Approval for development projects should only be granted after proven and implemented coastal protection measures.

3.5 It is necessary to establish regulations for prevention and protection for coastal defense structures following the principles outlined in (Zhachev N., Rafailov R., 2021) and to establish an engineering risk management system similar to the one proposed by (Rafailov R., 2020).

References:

[1]. Mocket I., Simm J., *Risk Levels in Coastal and River Engineerimg: a Guidance for Design*, Thomas Telford, 2018

[2]. Rafailov R., Stability of Road Embankments on Weak Soils, Geospatial Technology in Mining and Earth Sciences/ Selected papers of the 2nd International Conference on Geo-Spatial Technologies and Earth Resources 2022 (GTER2022 · GTER2022 Volume1) - Hanoi University of Mining and Geology (HUMG), Vietnam, 13-14 October 2022

[3]. Rafailov R.,“ System for risk management, prevention and protection on road infrastructure slopes“, Journal of Geology and Mineral Resources issue 1/2020, ISSN: 1310-2265

[4]. Zhechev N., Rafailov R., „Prevention and Protection of Road Infrastructure Slopes“, Journal of Geology and Mineral Resources, issue1-3, page 9, 2021, ,ISSN: 1310-2265

[5]. Zhechev N., „Socio-environmental Problems Caused by Landslides and Rockslides“, Journal of Geology and Mineral Resources, issue 4-5, page 9, 2021, ISSN: 1310-2265

Analysis of a two pile frame for a horizontal force

Iliyan Markov[1]

[1]Chief Assistant Prof. Dr. Eng., Department of Geotechnics, University of Architecture,Civil Engineering and Geodesy (UACEG), 1 Hristo Simirnenski Blvd., Sofia, Bulgaria

imarkov_fte@uacg.bg

Abstract. The article analyzes a pile frame composed of two bored piles with a diameter of 40 cm. The frame is subjected to a horizontal force. Calculations in three different method are conducted – subgrade reaction approach, 2D plaxis and 3D plaxis, after which a comparison was made between the results of the individual methods. The aim is to see under what conditions a subgrade reaction method is suitable for use without moving away from sure side. Results were also obtained for the distribution of the value of the spring constant by depth, as well as values for the ratio between the resistance of the second and first pile. A proposal has been made for obtaining a p-curve that can be applied to any of the linear approximation methods.

1. Introduction

Many methods have been developed in the literature for calculating a horizontal force pile, but unfortunately there is not one that is absolutely reliable. A conditional division of all methods can be made, dividing them into several groups [1]: (a) ULS - the ultimate limit state approach , (b) SRA - the subgrade reaction approach, (c) the continuum method, and (d) the finite element method (FEM). The methods of ultimate limit design [1] are represented by Blum approach, Brinch Hansen's Method, Broms Method in wich there is a static equation investigating the moment of limit state. The Subgrade Reaction Aproach (SRA) starts with Winkler's hypothesis. In this hypothesis, the soil reaction is replaced by spring constants that resist the vertical load. Later, rotating the beam to 90°, this method is used for a pile loaded with a horizontal force. A subsequent development of the method is the p-y method, in which the non-linear behavior of the soil is also taken into account, and the spring constants are now assumed to be non-linear. The condition of modeling the soil with linear or non-linear springs in SRA has been replaced in continuum approach [1] with a more realistic model. The soil is replaced by non-linear, linear and elastic elements. The behavior is represented as deflection around the vertical pile. Various models are presented: Simplified Continuum models (Energy and Variational Calculus Method, Modified Reissner Models), Boundary element approach [1]. The method of finite elements is used in all fields of mechanical modeling and it use the computing power to solve matrix equations. It is considered to be the most accurate method. The purpose of this development is to compare several models and to find a methodology to facilitate finding the correct springs value for subgrade reaction method

2. Guidelines influencing behaviour of piles, subjected to horizontal force

2.1. Difference in.soil.resistance between a one pile and a pile, which is part from a group

A single pile, subjected to a horizontal (lateral) force and a pile in a group of piles also subjected to a horizontal force have different behavior. The difference consists in reducing the bearing resistance of a pile , which is part from a group, which is called the shadow effect, seen in Fig.1. The resistance of the soil decreases with e

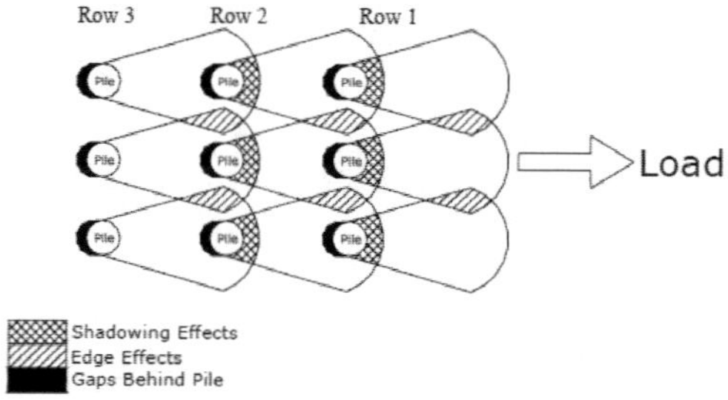

Figure 1. Edge effect and shadow on a pile group, loaded with a horizontal force [2]

There are a series of studies that suggest a different reduction factor in the second, third and each subsequent row. Snyder [3] introduces p_{mult} factor, which is lateral resistance ratio of pile in pile group and single pile, acc. to expression 1 and Fig.2. Brown[4] offers p_{mult}=0,8/0,4/0,3 for first, second, and third row respectively for free head. McVay [5] offers p_{mult}=0,8/0,45/0,3 for fixed head and medium dense sand (Dr=55%), and p_{mult}=0,65/0,45/0,35 for fixed head and medium dense sand (Dr=33%). Rolins[4] reports p_{mult}=0,8/0,45/0,3 for soft to medium clays.

$$p_{mult} = \frac{P_{gr}}{P} < 1,0 \quad (1)$$

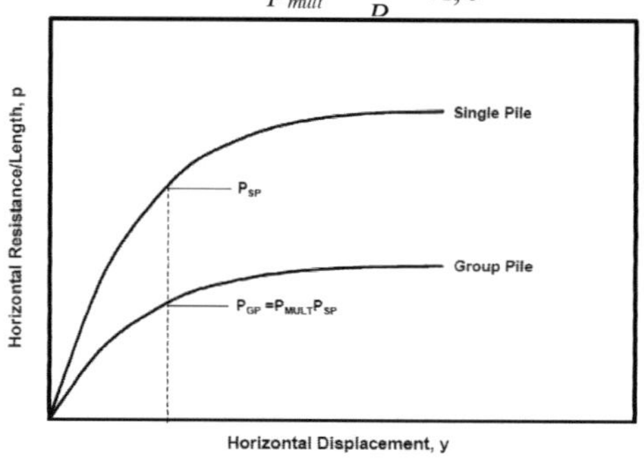

Figure 2. Plot for a single pile and random pile from a group [3]

2.2. Influence of distance between piles

Aforementioned authors do not account for pile spacing, which would also have an effect. FHWA [7] proposes to adopt a different reduction coefficient for different distances, according to the Table. 1

Table 1. Laterally load pile. Different value of P_{mult} for different row spacing.

Distance between rows	First Row	Second Row	Third and next rows
2D	0,80	0,45	0,35
3D	0,90	0,65	0,55
5D	1,00	0,85	0,75

Prakash in 1962, [9] suggest reduction factor to increase linear from 0.25 to 1 respectively for pile distance from 3D to 8D. p_{mult}=0.25 is too strict requirement and should be used for preliminary calculation. Distance between piles 8D is good solution for laterally load piles but if there is also a vertical force or bending moment it is too uneconomical.

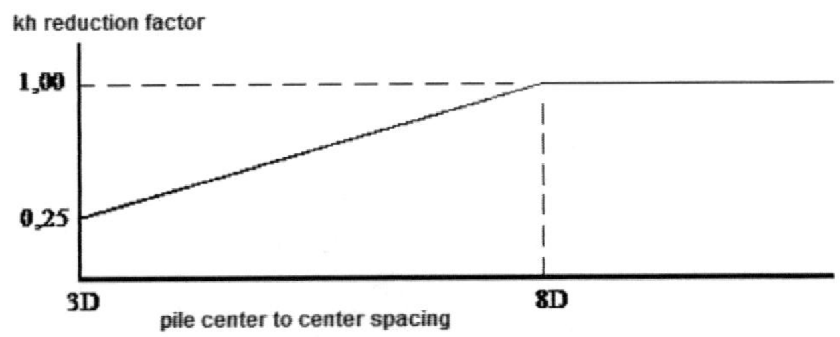

Figure 3. Group pile reduction proposed by Prakash in 1962, [9]

With a frequent change in the direction of loading, cyclic loading is observed and the last row becomes the first. In this case, to avoid a complex calculation procedure, a value of 0.5 p-multiplier may be used. Rollins [8] summarise results from researchers in period (1986-2003) as can be generalized for p_{mult}:

- first row $p_{mult} = (0.6\text{-}0.98)$
- second row $p_{mult} = (0.38\text{-}0.95)$
- third row $p_{mult} = (0.43\text{-}0.85)$

2.3. The p-y aproach

This method considers the ration between the soil resistance (p), and the transversal deflection of vertical pile (y). The representation of p-y relation in certain depth is named p-y curve. This model, used to analyze laterally loaded pile response is a refine modification of Winkler's basic model. The soil pressure per unit pile length is (p), and (y) is the displacement transverse to the pile axis. The main difference between p-y and Winkler model is that soils stiffness, represented by ratio p/y in Winkler model is linear, while in p-y method is non-linear. In p-y for particular depth is accepted different and not linear ratio p/y.

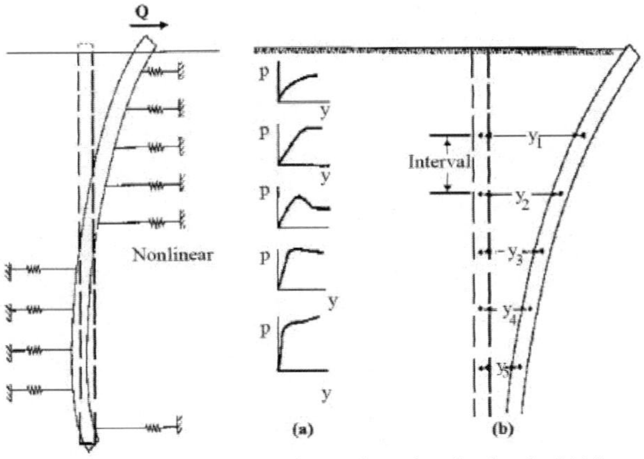

Figure 4. Different p-y plots, changing in depth, [10]

3. Geotechnical and structural parameters

The pile group is set on a one layer soil, according to Fig. 5 is the subject of the present study. The piles are two with a diameter of D400mm, united with a pile cap beam. The axial distance between them is 880, and the bearing depth is 800 cm. The horizontal force is applied at the level of the over-pile beam, which is 70 cm above the pile head.

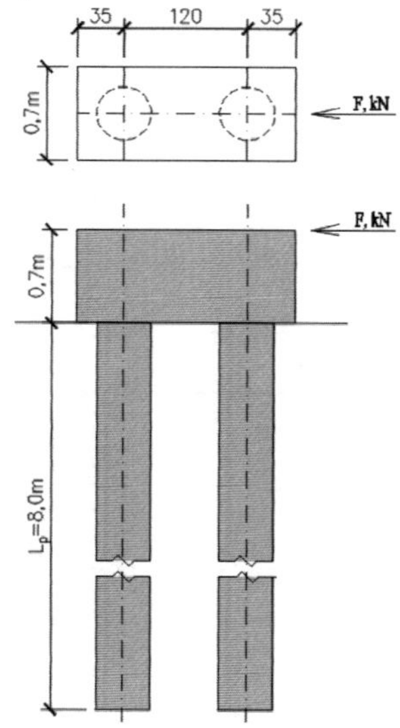

Figure 5. Soil Layers and pile group

- γ=15.5 kN/m^3 – volume weight,
- φ=32° - internal friction angle,
- c=3kPa – cohesion,
- E_{50}^{ref} – 30E3 kN/m^2
- E_{oed}^{ref} – 30E3 kN/m^2
- E_{oed}^{ref} – 90E3 kN/m^2

Subgrade reaction approach is one of the methods that can be used for preliminary calculation. Of course, it is not possible to replace the soil with a single number, but the purpose of this study is to provide methods to calculate preliminary values of displacement, bending moments and shear forces in the piles.

$$k_h = \frac{p}{w} \tag{2}$$

where:
p=resistance of soil per unit area (F/L^2)
k_h = coefficient of subgrade reaction (F/L^3)
w=beam settlement
The spring stiffness is calculate according to formula 3, [11]

$$k_{x,i}(z) = k_{h,i}.b_i.z_i.\overline{D},\ kN/m \tag{3}$$

where:
$k_{h,i}$, kN/m^4 - proportionality factor
$\overline{D} = 1.5d + 0.5m = 1,5.0,4 + 1 = 1,6m$ - effective width of the pile
b_i=1m

Table 2. Calculating spring stiffnesses

Spring stiffnes kN/m	Z, m	$k_{h.i,}$ kN/m^4	b_i, m	$\overline{D}$, m	$k_{x,1,i}$ kN/m	$k_{x,2,i}$ kN/m
$S_{st,1}$	1.0	9500	1	0,8	15200	10640
$S_{st,2}$	2,0	9500	1	0,8	30400	21280
$S_{st,3}$	3,0	9500	1	0,8	45600	31920
$S_{st,4}$	4,0	9500	1	0,8	60800	42560
$S_{st,5}$	5,0	9500	1	0,8	76000	53200
$S_{st,6}$	6,0	9500	1	0,8	91200	63840
$S_{st,7}$	7,0	9500	1	0,8	106400	74480
$S_{st,8}$	8,0	9500	0.5	0,8	60800	42560

With calculated in Table.3 spring constants, the frame was solved in the software Tower v 8.4. The diagrams of displacement, bending moments and shear forces are shown in Fig.6

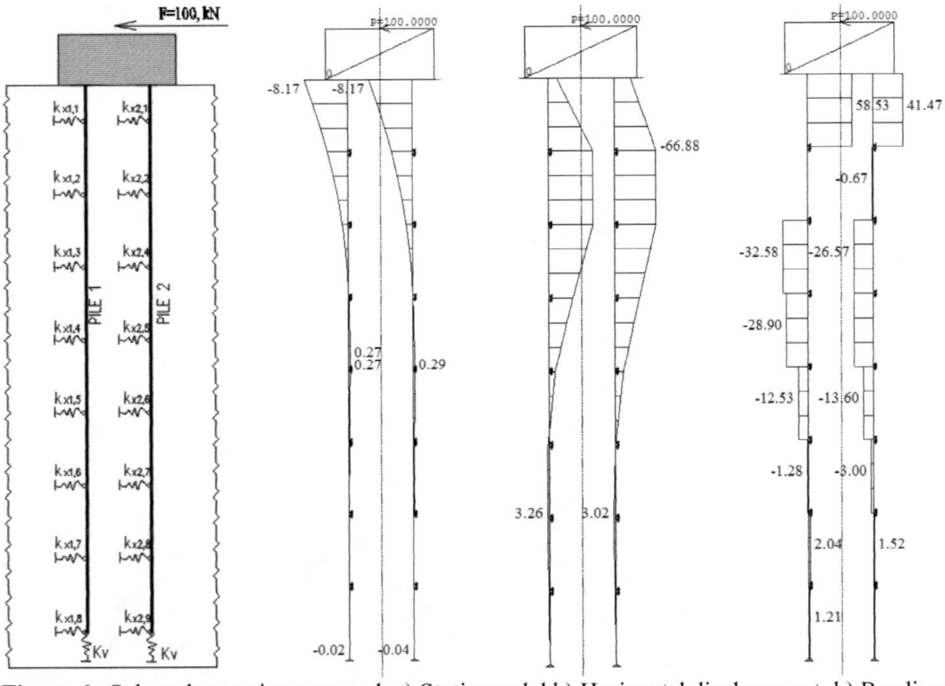

Figure 6. Subgrade reaction approach a) Static model b) Horizontal displacement b) Bending moment distribution d) Shear force distribution

4. 2D FEM method

In this model piles and pile cap are modelled as linear frame elements. Piles are modeled as embedded beam row with D=40 cm, E=31.5E6 kN/m^2, predifined beam type and $L_{spacing}$=1m. Capping beam is presented in model with plate element, relevant to concrete material and dimensions as in Fig.2 A very basic, but sufficient for preliminary calculations method is to obtain k_h in depth of pile. This is possible by finding the slope of the shear force diagram (first derivate), which is actually the load p from formula 1. Here, for convenience, a unit force F=100kN is applied. Obtained values of should be multiplied for the real value of horizontal force F. The other results is ratio between $k_{h,2}$ and $k_{h,1}$, presented in Formula 4

$$\eta_{mult} = \frac{K_{h2}}{K_{h1}} < 1,0 \tag{4}$$

To eliminate the influence of other parameters, vertical force and bending moment are not included.

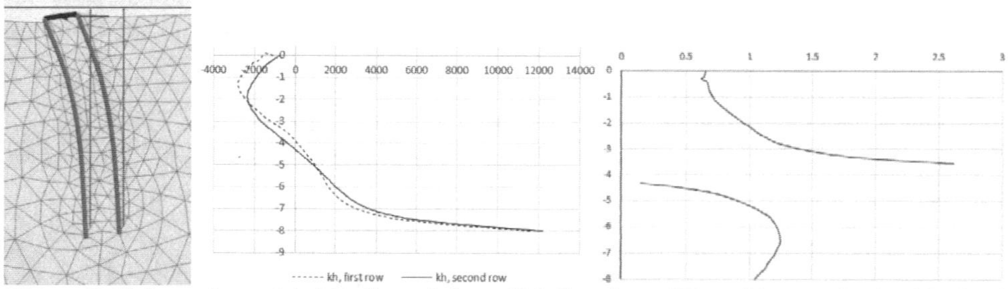

Figure 7. Plain strain model a) Deformed shape b)Soil spring stiffness b) $\eta_{mult,}$ obtained for load F=100kN

Thus obtained values of spring constants should be entered in any program for calculating building structures. The special thing is that they are valid only for the selected type of pile frame and the selected type of construction. The approach is linear and does not limit the growth of the spring constants, which should also be observed

5. 3D model

In order to obtain the behavior of k_h from Formula 1 a three-dimensional model was compiled in which the spatial effect of the resistance in front of the first and second pile is taken into account, monitoring the change of k_h

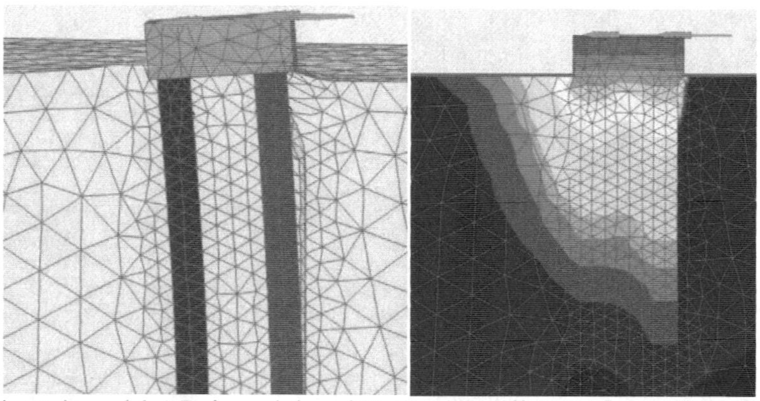

Figure 8. Plain strain model a) Deformed shape b)Soil spring stiffness and $p_{mult,}$ obtained for load

5.1. P-y curves, obtained from 3D model

From model are plotted p-y diagrams for some nodes in depth for first and second row of piles.

In order to obtain the behavior of k_h from Formula 1 a three-dimensional model was compiled in which the spatial effect of the resistance in front of the first and second row of piles is taken into account, monitoring the change of k_h

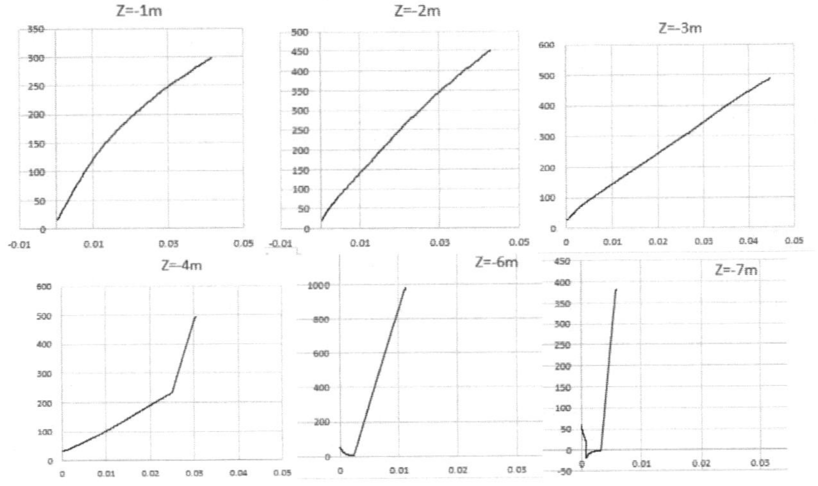

Figure 9. The plot of p-y for first pile, achieved from 3D-model.

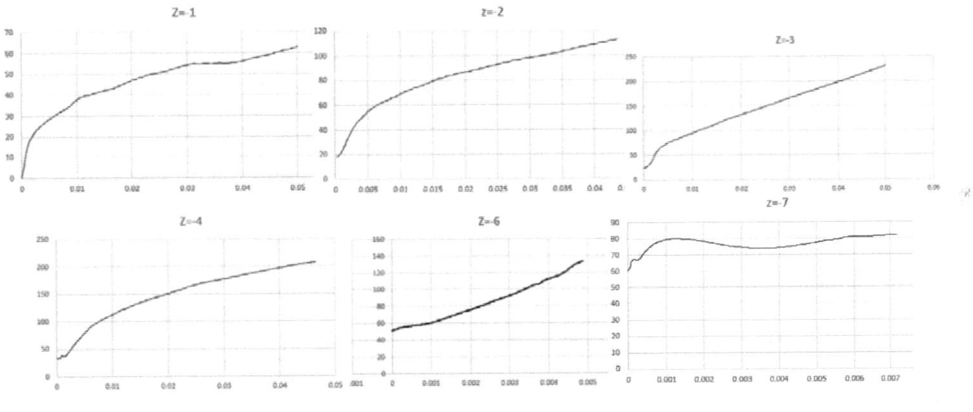

Figure 10. The plot of p-y for second pile, achieved from 3D-model.

5.2. P_{mult} factor

A practical result of this model is the calculated value of p_{mult}, according to expression 1. The value for different depth is: $P_{mult}(-1m)=0.7$; $P_{mult}(-2m)=0.87$; $P_{mult}(-3m)=0.95$; $P_{mult}(-4m)=1,0$; $P_{mult}(-6m)=1,0$; $P_{mult}(-7m)=1,0$;

6. Comparison of the results

Three models have been made from. The result is comparison of displacement, shear force and bending moments.

Table 3. Results of the three models for horizontal force F=100kN

Spring stiffnes kN/m	$x_{pc,}$ mm	$M_{max,1}$ kN/m	$M_{max,2,}$ kN/m	$Q_{pc1,}$ kN	$Q_{pc2,}$ kN
2D Tower model	8.17	72.03	66.88	58.53	41.47
2D Plaxis Model	14.65	35.03	34.32	51.37	42.46
3D Plaxis Model	28.34	60.94	51.4	33.48	31.9

Here it can be concluded that from the model in which the soil is replaced by spring constants, the forces are in favor of security while the displacements are underestimated. $M_{max,1}$ is abaut 20% bigger than $M_{max,2.}$ $M_{max,1}$ is also about 20% bigger then $Q_{max,2}$.

7. Conclusions

In the three models made, it can be seen that there is no unequivocal answer for the value of bending moments, shear forces, and displacements, and can be concluded:

- The subgrade reaction approach is on sure side about the bending moments and shear force, but is not reliable for displacement,
- The soil resistance is not growing linearly in depth,
- The most accurate method for achieving spring constant value is calculate p-y curve from Plaxis 3D and input them in any structural software that use subgrade reaction approach,
- The ratio between resistance of second and first pile is not constant

References

[1] Moussa A and Christou P 2018 *The Evolution of Analysis Methods for Laterally Loaded Piles Through Time*
[2] Walsh J M 2005 *Full-Scale Lateral Load Test of a 3x5 Pile Group in Sand* (Thesis, Brigham Young University)
[3] Snyder J L 2004 *Full-Scale Lateral Load Test of a 3x5 Pile Group in Soft Clays and Silts* (Thesis, Brigham Young University)
[4] Brown D A Morrison C and Reese L C 1988 *Lateral Load Behavior of Pile Group in Sand* (Journal of Geotechnical Engineering, Vol. 114, No. 11, November.)
[5] Mitchell J K Baxter C D P and Munson T.C 1995 *Performance of Improved Ground During Earthquakes‖, Soil Improvement for Liquefaction Hazards Mitigation* (Geotech. Special Pub. No. 49, ASCE, 1-36.)
[6] Rollins, K. M., Weaver, T. J., and Peterson, K. T. 1997 *Statnamic Lateral Load Testing of a Full-ScaleFixed-Head Pile Group,* (Proceedings of the National Seismic Conference on Bridges and Highways, Sacramento, California, Federal Highway Administration and California Department of Transportation, July, pp. 267 - 282.)
[7] Kavazanjian E et al. 2011 *LRFD Seismic Analysis and Design of Transportation Geotechnical Features and Structural Foundations* (Publication No. FHWA-NHI-11-032 Federal Highway Administration GEC No. 3)
[8] Rollins K M 2003 *Response, analysis and design of pile groups subjected to static and dynamic lateral loads (*Utah Department of Transportation Research and Development Division)
[9] Prakash A 1962 *Behavior of pile groups subjected to lateral loads, (Thesis, University of Ilinois)*
[10] Christensen D S 2006 *Full Scale Static Lateral Load Test of a 9 Pile Group in Sand* (Thesis, Brigham Young University)
[11] Ilov. G et al 2011 *Handbook of Geotechnics*

Advanced technologies for landslide monitoring

N. Milev [1], A. Totsev [2], M. Angelova [3]

Abstract. The article reviews modern landslide monitoring methods. The purpose of the observation method in the design and control of the behavior of geotechnical structures and its role in creating chances for cost savings or pre-emptively averting construction risks is a driving force in the context of landslides. This approach is established within Eurocode 7 as a fundamental tenet of geotechnical design, serving as a valid validation method for limit states. Monitoring of large areas is common in everyday engineering practice which makes the installation of transducers in multiple points very difficult. Due to this reason the use of so-called contactless approaches is widely accepted. A distinguishable representor of that method is DInSAR – technique for successful displacements measurement by means of SAR data which is economically feasible and capable of detecting classic failure modes. The article presents and compares methods such as: LiDAR, Infrared sensors, Fiber oprics, Shape array accelerometers, In-place inclinometers, and others. Despite the development of technologies, the application of monitoring in Bulgaria is not enough used.

1. DInSAR

Observing the real-time actions of landslides holds pivotal importance in evaluating and foreseeing the potential disaster risks they pose. Nonetheless, due to the typically vast scope of the targeted regions, consistent monitoring has often been hindered by technical complexities and financial constraints. Synthetic Aperture Radar (SAR), a sophisticated radar apparatus situated on an artificial satellite, emerges as a noteworthy solution. By emitting pulse waves towards the Earth's surface and capturing their reflections, SAR generates high-resolution data. A technique known as Differential Interferometric SAR (DInSAR) leverages SAR information to identify surface displacements, exhibiting promising attributes for proficient, impactful, and cost-efficient monitoring of extensive landslide behaviors. As demonstrated on figure 1, it uses two scenes of SAR data and the displacement obtained by that technique is one-dimensional along the LOS (line of sight), i.e., extension and compression – (Ferretti, 2014). One notable benefit offered by Differential Interferometric SAR (DInSAR) is its ability to oversee expansive regions without requiring ground-level sensor installations. Its unique capabilities have made it a valuable tool in various geotechnical contexts, providing critical insights into geological and environmental processes.

[1] Nikolay Milev, Chief Assist. Prof. Dr. Eng., UACEG, 1000, Sofia, Hristo Smirnenski Blvd 1, n.milev@yoda-bg.com
[2] Andrey Totsev, Prof. Dr. Eng., UACEG, 1000, Sofia, Hristo Smirnenski Blvd 1, atocev@gmail.com
[3] Melanie Angelova, Student TE, UACEG, melanierick.angelova@gmail.com

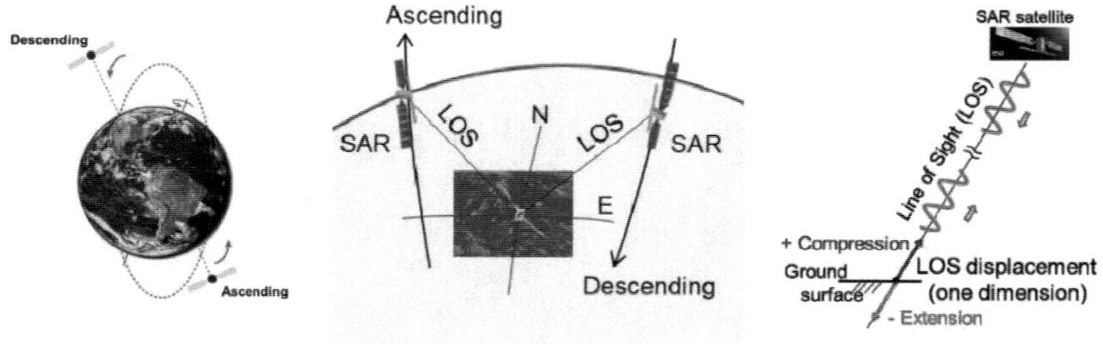

Figure 1. a) SAR satellite orbits b) Image acquisition from ascending and descending orbits c) Line of Sight (LOS) and LOS displacement (Corsa 2022)

Landslides pose significant hazards to human lives and infrastructure, making early detection and risk assessment essential. DInSAR's ability to measure millimeter to centimeter-scale ground movements aids in the identification of potential landslide-prone areas and the assessment of landslide dynamics as shown on figure 2 – (Yamaguchi,2021). By continuously monitoring these areas, authorities and researchers can better understand the factors contributing to landslide occurrences, such as rainfall, geological conditions, and human activities, and develop more effective mitigation strategies. In landslide monitoring, DInSAR offers several advantages over traditional methods. Firstly, it provides continuous and real-time monitoring, enabling early detection of landslide activities. Unlike conventional ground-based surveys, which can be time-consuming and intermittent, DInSAR allows for continuous surveillance, ensuring that any ongoing or evolving landslide movements are promptly detected. Secondly, DInSAR's wide area coverage facilitates comprehensive observation of landslide-prone regions. It can cover extensive areas, providing a holistic view of potential landslide hazards. This wide spatial coverage is particularly advantageous in monitoring regions with complex terrain and multiple potential landslide sites. Moreover, DInSAR is unaffected by weather conditions, offering consistent data acquisition regardless of cloud cover or darkness and its high precision in measuring ground surface displacements is instrumental in assessing landslide risks. Unlike optical satellite imagery, which can be hindered by atmospheric interference, DInSAR's radar signals penetrate through clouds, providing reliable measurements under all weather conditions.

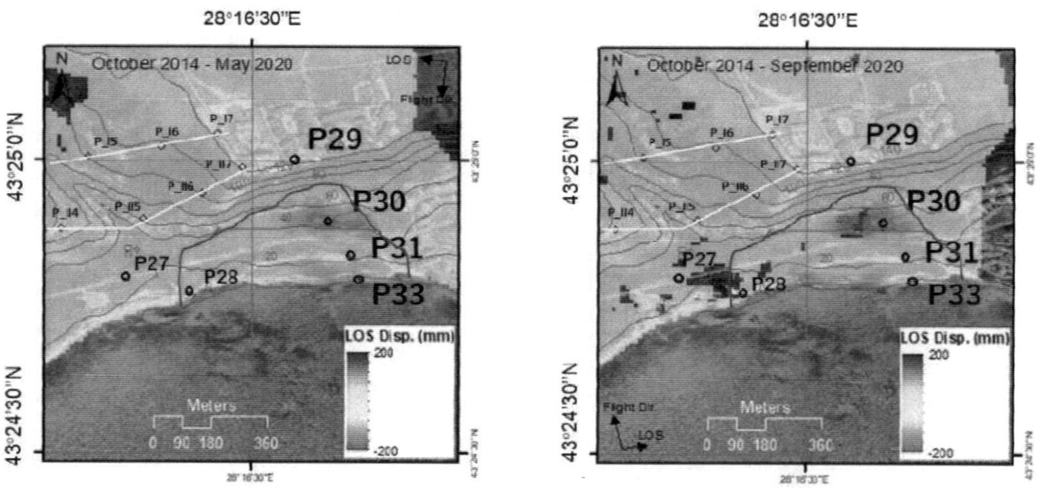

Figure 2. LOS displacement distributions around eastern end of a landslide in Bulgaria (Topola village)

IOP Conf. Series: Materials Science and Engineering 1297 (2023) 012008 doi:10.1088/1757-899X/1297/1/012008

DInSAR plays a crucial role in monitoring ground subsidence, which is the gradual sinking or settling of the Earth's surface. Ground subsidence arises from a range of influences, encompassing natural occurrences like sediment compaction and anthropogenic actions such as mining and extracting groundwater. DInSAR's precision in measuring subsidence rates allows for the identification of areas experiencing subsidence, enabling engineers and urban planners to assess potential risks to infrastructure, such as buildings, roads, and pipelines. With this information, appropriate measures can be taken to mitigate subsidence-induced damages and preserve the integrity of critical structures.

Another significant geotechnical application of DInSAR is the monitoring of volcanic deformation – (Corsa, 2022). Volcanoes are complex geological features that can exhibit various forms of deformation prior to, during, and after volcanic eruptions. Through the utilization of Differential Interferometric SAR (DInSAR), scientists and geologists gain the capacity to identify nuanced alterations in the Earth's exterior. This encompasses phenomena like uplift or subsidence that are linked with volcanic operations. This information is invaluable for assessing volcanic behavior, forecasting potential eruption events, and understanding the underlying magmatic processes. By closely monitoring volcanic deformation, authorities can make informed decisions regarding hazard assessments, evacuation planning, and volcanic risk management, thus safeguarding communities living in volcanic regions.

Furthermore, DInSAR's applications extend to the monitoring of critical infrastructure, including dams, bridges, and pipelines – (De Corso, 2020). Ensuring the stability and integrity of such infrastructure is paramount to prevent catastrophic failures. DInSAR offers a cost-effective and wide-area monitoring solution that complements traditional surveying methods. By continuously tracking ground movements around infrastructure sites, engineers can detect any signs of deformation or settling early on and take timely remedial actions, enhancing public safety and asset protection.

In the oil and gas industry, DInSAR has proven beneficial in monitoring ground movements related to reservoir activities. Oil and gas reservoirs undergo complex changes in subsurface pressure during extraction and production. DInSAR allows for the observation of ground surface deformations resulting from these pressure changes, providing valuable data for reservoir management and environmental impact assessment. This information assists in optimizing production strategies and mitigating any potential environmental effects associated with hydrocarbon extraction.

To leverage the full potential of DInSAR in geotechnical monitoring, it is essential to integrate and analyze the data in conjunction with other geospatial and geological information. Combining DInSAR data with ground-based monitoring techniques, geological surveys, and geotechnical models enhances the accuracy and reliability of the results, thereby facilitating more informed decision-making.

In conclusion, DInSAR has emerged as a versatile and indispensable technology for geotechnical monitoring. Its capability to detect ground movements over large areas and with high temporal resolution opens new horizons for understanding geological processes and assessing risks. From landslide monitoring and ground subsidence assessment to volcanic deformation tracking and critical infrastructure surveillance, DInSAR plays a pivotal role in ensuring human safety, protecting infrastructure, and optimizing resource management. As technology continues to advance, DInSAR's potential for geotechnical applications is poised to grow, making it an increasingly invaluable asset in addressing the challenges of an ever-changing Earth.

2. LiDAR

Light Detection and Ranging (LiDAR), an innovative technology for remote sensing, has risen as a revolutionary instrument in the surveillance of landslides. This technique utilizes lasers to gauge distances by projecting laser pulses towards the Earth's surface and gauging the duration it requires for the laser to rebound subsequent to encountering an object or the terrain. By collecting millions of laser points, LiDAR creates a detailed and accurate 3D point cloud representation of the surveyed area. This high-resolution data enables the creation of highly precise digital elevation models (DEMs) and detailed terrain analysis – figure 3.

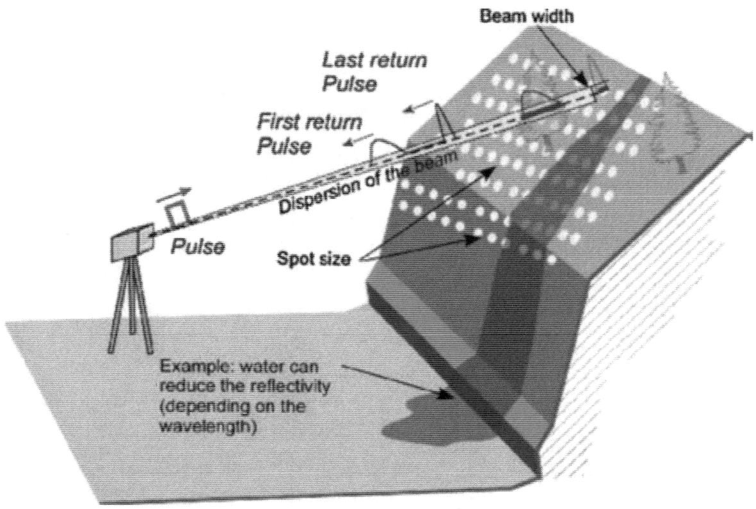

Figure 3. Illustrating the process of obtaining laser scanner data, exemplified through the use of LiDAR principles (Jaboyedoff, 2012)

One of the key advantages of LiDAR is its exceptional spatial resolution. It can capture terrain and objects on the Earth's surface with centimeter-level accuracy, providing detailed information on the landscape's topography and surface features. This level of precision is invaluable for identifying potential landslide-prone areas and assessing their vulnerability. Moreover, LiDAR's ability to generate precise 3D models of the terrain allows for a comprehensive understanding of the landscape's morphology. This detailed mapping aids in detecting subtle ground movements, potential landslide scarps, and other indicators of landslide activity. LiDAR technology can be deployed for continuous monitoring, facilitating real-time observations of the landscape's topographic changes. By comparing repeated LiDAR scans over time, geotechnical experts can identify ongoing landslide movements and assess their dynamics more accurately.

The precise and real-time monitoring capabilities of LiDAR are crucial in developing early warning systems for landslides. Continuous LiDAR scans can detect even subtle changes in the landscape, allowing for timely alerts and prompt responses to potential landslide threats. Furthermore, by comparing multiple LiDAR scans, geotechnical experts can analyze the speed and direction of landslide movements. This insight helps in predicting the extent of the landslide's reach and potential impact on critical infrastructure and communities.

3. Fiber optic sensing (FOS)

One of the most advance techniques for geotechnical monitoring is fiber optic sensing (figure 4). Using this new technology, the temperature, strain, and vibration can be measured of a single fiber continuously of different intervals, like for example 5, 10 or 20 cm, along the length of the cable. Fiber optic cable can be with length from few meters to several kilometers and allows large databes collection. Prof. Soga's (Soga& Luo 2018) use very good comparison to describe the fiber optic cables as a "civil infrastructure nerve system". If an optical fiber cable is attached to geotechnical structure – the fiber optic cable provide thousands of "strain gauges", "thermo-couples", or "accelerometers".

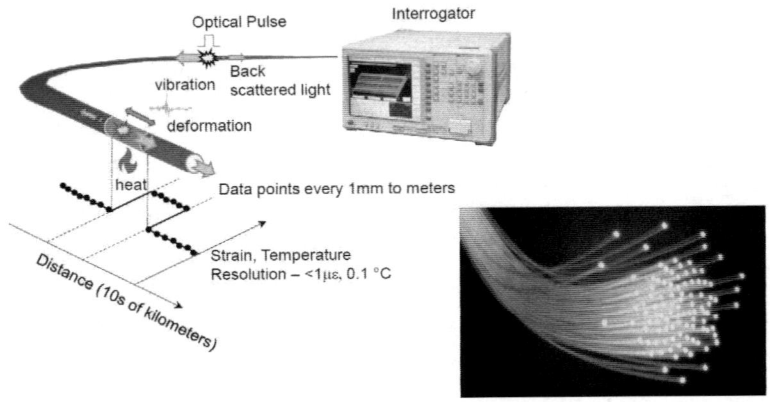

Figure 4. Fiber optic sensing (Soga etc. 2018)

The principle of operation of the optical fiber is to guide the light beam along the axis by total internal reflection. Energy is transmitted along the optical waveguide at received wavelengths in the visible part of the electromagnetic spectrum or in the infrared part of the electromagnetic spectrum (Agrawal 2006). The cross-section of an optical fiber is very small – about 390 μm. It can be divided of three components - central core, cladding and buffer figure 5 (a). Based on Brillouin scattering the technology allows the strain or temperature to be measured along a cable connected to a spectrum analyzer.

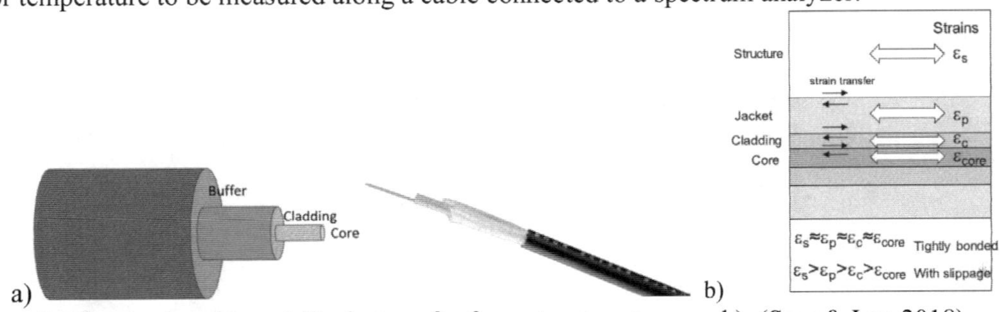

Figure 5. Fiber optic cable - a) Strain transfer from structure to core; b) (Soga& Luo 2018)

In geotechnical practice the most common use of fiber optic sensing is for determination of the strains. Using optical fibers in a geotechnical structure, strain is transmitted from it to the core of the optical fiber by shearing through the bonded surfaces between the different materials of the structure as shown in Figure 5 (b). Fiber optic cable can be used for monitoring of piles, for tunnel monitoring, for anchors, for slope monitoring etc. (figure 6). The future development of this technique will find great application in geotechnical monitoring.

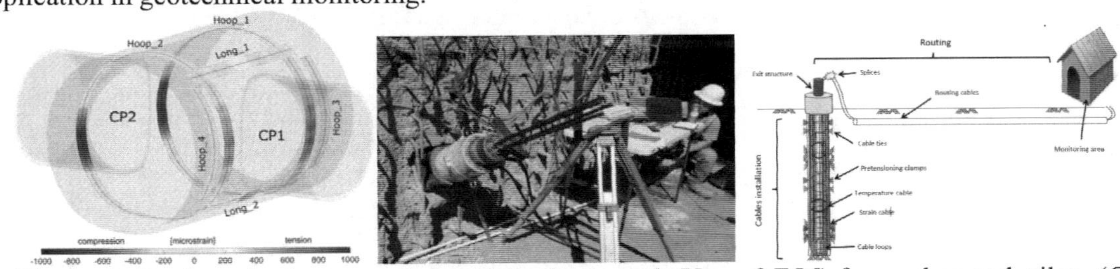

Figure 6. Total strain increase observed in the tunnel. Use of FOS for anchor and piles. (Geo monitoring course, Berkley 2023)

IOP Conf. Series: Materials Science and Engineering 1297 (2023) 012008 doi:10.1088/1757-899X/1297/1/012008

4. Shape array accelerometers (SAA)

Shape array accelerometers or Shape Accel Array accelerometers are a series of arrays of sensors (microprocessors), which mainly determine the deformations and, in particular, the vibrations. It consists of row of solid segments with different length (form 200 mm to 500 mm). The diameter is 25 mm. Every part (segment) has three micros electro mechanical system accelerometers (figure 7).

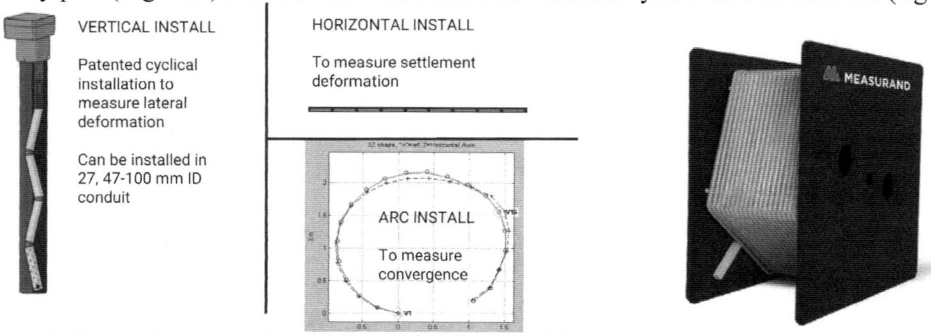

Figure 7. Shape Array accelerometers (K. Johnson 2023)

Individual elements are connected to each other with joints (flexible). These joints allow movement in all directions except for twisting. One processor including a temperature sensor is placed on every eight segments. Together with microprocessors in array collect data and send the information through wire connect with the SAA. The length of the SAA can be 100 m and more (Bradley 2022).

Very interesting example of use of SAA on landslide is presented by Dasenbrok (Dasenbrok 2010). During the monitoring one of the landslides have collapsed. On figure 8 are presented some data. This project have been one of the first use of SAA for landslides and one of the conclusions is that in comparison with traditional system (classical inclinometers), the SAA systems provided a significantly improved level of detail, due principally to their sampling frequency and ductility.

Figure 8. Example with use of SAA on landslide monitoring (Dasenbrok 2010)

5. Infrared termography (IRT)

Infrared thermography is a new technology that can also be applied in landslide reinforcement. It also serves for mapping by describing different movements of the slope. IRT can be applied from ground and air platforms. Usually IRT is applied in conjunction with other geomatc methods - TLS and GPS to monitor the behavior of landslides - new one and existings. The principle of operation of IRT is based on the principle that objects with a temperature greater than absolute zero cause electromagnetic radiation that varies in the limits of the infrared band (Fordella 2017). Figure 9 shows the infrared region as well as the spectral energy distribution.

According to (Lillesand 2008), all thermal radiation (thermal infrared energy) at the earth's surface consists of IR of various wavelengths, with the IR radiation responsible for thermal radiation varying from 0.70 μm to about 1000 μm.

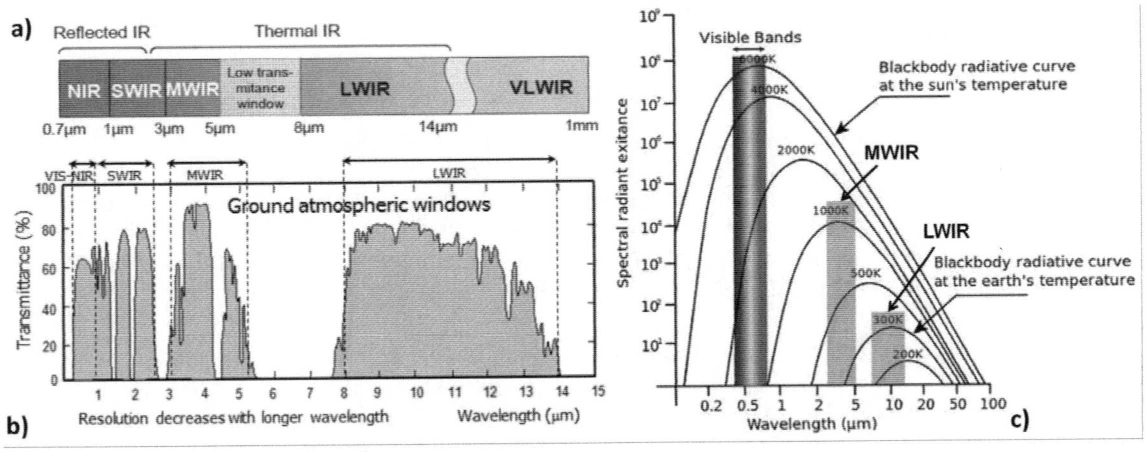

Figure 9. The electromagnetic spectrum in the infrared range (a) and the spectral region in the range from 0 to 15 m (b); black bodies spectral distribution of energy showing the infrared zones characteristic of the sensors of the latest generations of thermal cameras (c) (Fordella 2017).

An example is presented on figure 10 (Dzekkos 2022) with infrared orthophotos of failure of the Sanford Dam. With thermal infrared cameras were prepared infrared imagery. Infrared imagery identifies the presence of a sub-drain system, geospatial assessment of encountered materials and geospatially variable moisture contents.

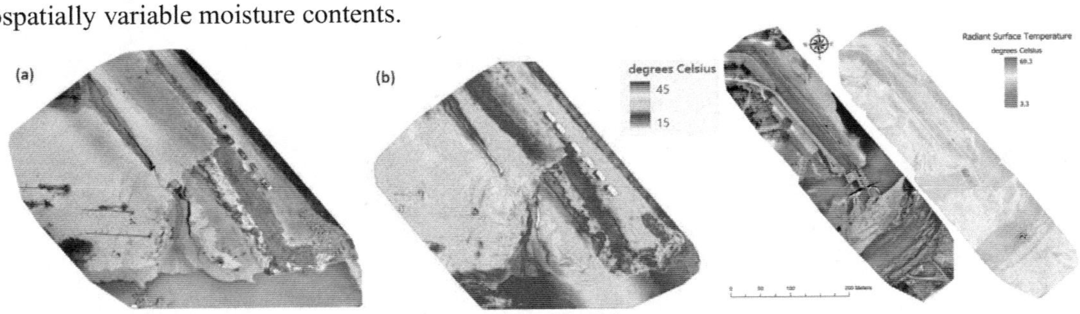

Figure 10. Optical and infrared imagery (Dzekkos 2022)

Another examples from Italy of thermal anomalies connected to slope stability phenomena are presented on the figure 11. - open fractures and water seepage in sandstone layers. By thermography using IR thermally calibrated cameras is infrared imaging received. The surface temperature is represented by means of a color scale. The higher temperatures are presented with the lighter colors and the colder temperatures with darker.

Figure 11. Thermal anomalies connected to slope stability phenomena - open fractures and water seepage in sandstone layers (Fordella 2017).

6. In-place inclinometers

The inclinometer in place is a geotechnical instrument used to monitor deformations in the ground or displacements in a given structure figure 12 (a). The device is part of a whole system, the so-called the inclinometer system - is a probe that is installed in a housing that protects it from deformations, as the inclinometer is installed and powered by a cable, which cable is placed in the probe (Timothy D,2008). Bonded together and suspended inside the inclinometer housings, the MEMS sensors are a major part of the system. The probe moves along the length of the housing with the help of the cable. It measures the hull deformation and transfers the data to the data acquisition system (memory).

Measurements are taken from bottom to top at 50 cm intervals, with the first reading ie. zero defines the initial shape of the hull. From the subsequent readings, a readout of the tilt along both axes (deformation) is obtained by measuring and extracting data. The accuracy of an inclinometer is about 6.7 millimeters at 30 meters. Systematic errors are obtained that affect the accuracy of the measurements due to errors in the probe assembly, as well as random errors that are related to the function of the sensors. (Hua-Fu Pei1, 2012). If the base of the inclinometer housing is considered stable, the slope data from the bottom is accumulated to create cumulative results. Otherwise, the upper part of the housing is considered as the reference point figure 12(a). The results show magnitude, speed, depth, type and direction of movement, which are made clear by the difference between the zero reading and subsequent readings, and are saved in data loggers. They are processed by software and presented graphically (diagrams) figure 12 (b), including cumulative and incremental measurements.

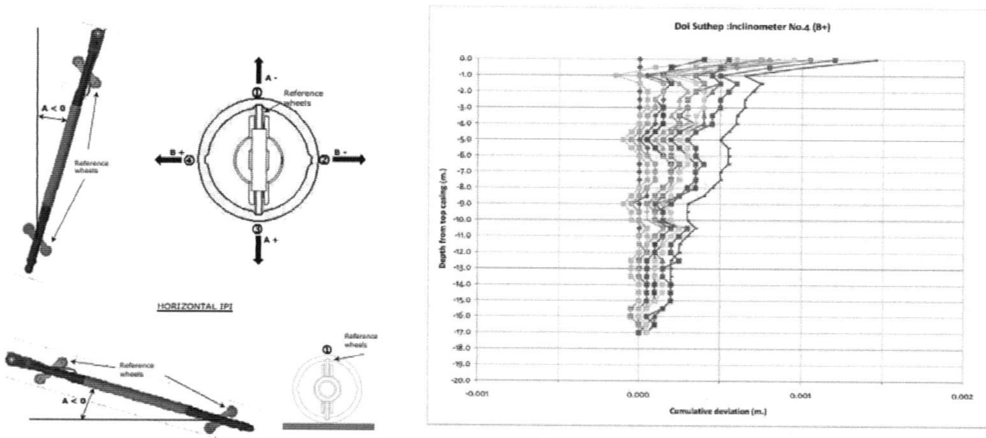

Figure 12. a) Installation of an inclinometer (SISGEO) b) Presentation of measurement data (Method Statement INCLINOMETER SYSTEM, Infratech ASTM Co.,Ltd)

The applications of inclinometers in geotechnical monitoring are extensive, and both vertical and horizontal inclinometers could be used. We observe them most often when measuring the stability of a given slope or steep section, landslides, monitoring of pile and diaphragm walls, excavations, deformations in embankments, they are also widespread in road, tunnel and bridge construction (figure 13).

IOP Conf. Series: Materials Science and Engineering 1297 (2023) 012008 doi:10.1088/1757-899X/1297/1/012008

Figure 13. MEMS Inclinometers in situ (MEMS Inclinometers on place , Sisgeo)

Inclinometers are geotechnical instruments that discuss different types of ranges in modern construction activity by determining and pre-confirming important deformations (Zhechev). They monitor slope stability by indicating potential damage or accelerated deformation. In conclusion, they mitigate as many risks as possible to lead to more successful projects.

Conclusion

The presented advanced techniques for monitoring of landslides are a powerful tool for preventing failure (Nedkov et al, 2014). They are a means of realizing significant savings in the design of landslide fortification facilities. The future design and development of landslides protection is unthinkable without use of geotechnical monitoring. The engineering practice recommend the simultaneous use of several methods to reduce the potential error.

References

[1] Bradley B and Garcia R., 2022 The use of shape accel array for monitoring utilities during urban tunnel drives

[2] Corsa, B., Barba-Sevilla, M., Tiampo, K., & Meertens, C. 2022. Integration of DInSAR time series and GNSS data for continuous volcanic deformation monitoring and eruption early warning applications. *Remote Sensing, 14*(3), 784.

[3] De Corso, T., Mignone, L., Sebastianelli, A., del Rosso, M. P., Yost, C., Ciampa, E., & Ullo, S. 2020. Application of DInSAR technique to high coherence satellite images for strategic infrastructure monitoring. In *IGARSS 2020-2020 IEEE International Geoscience and Remote Sensing Symposium* (pp. 4235-4238). IEEE.

[4] Dasenbrock D., 2010 *Automated Landslide Instrumentation Programs on US Route 2 in* Crookston

[5] Ferretti, A., 2014. *Satellite InSAR data: reservoir monitoring from space*. EAGE Publication. The Netherlands.

[6] Govind A., 2006, *Nonlinear fiber optics.*

[7] *Inclinometer : Types , How It Works, & Uses* , Encardio , December 03, 2019

[8] Jaboyedoff, M., Oppikofer, T., Abellán, A., Derron, M. H., Loye, A., Metzger, R., & Pedrazzini, A. 2012. Use of LIDAR in landslide investigations: a review. *Natural hazards, 61*, 5-28.

[9] Johnson K, 2023 *Shape array accelerometers, Geo monitoring course,* Berkley

[10] Lillesand, T.M.; Kiefer, R.W.; Chipman, J.W. Remote Sensing and Image Interpretation, 6th ed.; John Wiley and Sons, Inc.: Hoboken, NJ, USA, 2008.

[11] *Monitoring of lateral displacements of a slope using a series of special fiber Bragg grating-based in-place inclinometers*, 17 January 2012 Hua-Fu Pei1, Jian-Hua Yin1, Hong-Hu Zhu2, Cheng-Yu

Hong1, Wei Jin3 and Dong-Sheng Xu1 1 Department of Civil and Structural Engineering, The Hong Kong Polytechnic University, Hung Hom, Kowloon, Hong Kong, People's Republic of China 2 School of Earth Sciences and Engineering, Nanjing University, Nanjing, Jiangsu Province

[12] Nedkov N., L.Hristchev, Ch. Kolev, K. Radlov, *Performance of special geotechnical works – deep mixing of soil with binder substance, First Scientific applied conference with international participation* “Project Management in Civil Engineering”, UASG, Sofia, 4-5 december 2014, Collection of papers, pages 288-293 (ISSN 2367 – 6752 (CD/DVD));

[13] Sisgeo, *MEMS Inclinometers on place.*

[14] Soga K. and Luo L., 2018, *Distributed fiber optics sensors for civil engineering infrastructure sensing, Journal of Structural Integrity and Maintenance*, VOL. 3, NO . 1, 1–21, 2018

[15] Soga K at all, Fiber-Optic Underground Sensor Networks, 2018

[16] Timothy D. Stark, University of Illinois at Urbana-Champaign, Landslide Inclinometers, Article in Landslides · August 2008

[17] W. Frodella at all, 2017 Landslide Mapping and Characterization through Infrared Thermography (IRT): Suggestions for a Methodological Approach from Some Case Studies.

[18] Yamaguchi, M., Yastika, P.E., Shimizu, N., Milev, N. & Vrkljan, I. 2021. Application of SBAS-DInSAR to monitoring landslides along the northern Black Sea coast in Bulgaria, *Proceedings of EUROCK-Torino 2021*, IOP Conf. Series: Earth and Environmental Science 833, 012151. doi:10.1088/1755-1315/833/1/012151

[19] Zekkos D. at all, *UAV-Enabled Coupled Infrared and Optical Characterization* of the May 19, 2020, *Edenville Dam Failure in Michigan, Geo Congress* 2022

[20] Zhechev N., Prevention of activated landslides through the human factor and the legislation for geo-protective construction/ 5-th International Multidisciplinary Scientific Geoconference /SGEM 2005/- Modern Management of Mine Producing, Geology and Environmental Protection, Bulgaria – Albena.

[21] Zhechev N., GLOBAL WARMING – A CONCEPT OF UPDATING OF DESIGN CRITERIA FOR GEOTECHNICAL STRUCTURES, 7-nt International Multidisciplinary Scientific Geoconference /SGEM 2007/- Modern Management of Mine Producing, Geology and Environmental Protection, Bulgaria – Albena.

IOP Conf. Series: Materials Science and Engineering 1297 (2023) 012009 doi:10.1088/1757-899X/1297/1/012009

Rapid detection of landslides mechanisms and assessment of their geometry and dimensions by means of a drone survey (UAV) after the 2023 Turkey-Syria earthquake

N. Milev[1], **T. Tobita**[2], **T. Kiyota**[3] **and M. Shiga**[4]

Abstract. The 2023 Turkey-Syria earthquakes (M_w=7.7 and M_w=7.6, respectively) had serious impact on human lives, infrastructure and buildings. Unfortunately, more than 15 million people have been affected and current estimation of financial loss exceeds 35 billion dollars. Furthermore, more than 550 M_w>4 aftershocks have been recorded since the catastrophic events of February 6th 2023. Such emergency situations require rapid actions for assessing the failure mechanism and scale of problematic landslides since immediate decisions should be taken in most extreme cases. Recent advances in technology allows the application of relatively simple methods for such emergency engineering estimation – a good example is the use of drones (UAV) which are easily available on market nowadays. Drone videos allow the subsequent measurement of important parameters of landslides, such as length, width, depth, failure plane inclination and volume of displaced soil. The presented paper demonstrates three case studies which are based on drone video recordings: two landslides (Islahiye and Tepehan, respectively) and a rockfall (Idili).

1. Introduction

In the aftermath of seismic events, landslides impose formidable challenges on disaster response and geotechnical monitoring teams. Timely and accurate assessment of affected areas is critical for effective disaster management and rescue operations. Embracing the democratization of technology, everyday drones have emerged as a powerful tool for urgent landslide monitoring after earthquakes, offering unprecedented capabilities that accelerate rescue efforts and safeguard human lives. This paper examines the pervasive impact of everyday drones in post-earthquake landslide monitoring, focusing on their accessibility, real-time imaging, geospatial capabilities, multi-spectral and thermal imaging, community involvement, and implications for disaster management.

[1] Nikolay Milev, Chief Assistant Prof., Department of Geotechnics, University of Architecture, Civil Engineering and Geodesy (UACEG), 1 Hristo Simirnenski Blvd., Sofia 1164, Bulgaria, milev_fte@uacg.bg
[2] Tetsuo Tobita, Professor, Kansai University, Department of Civil, Environmental and Systems Engineering, 3-3-35, Yamate-cho, Suita-shi, Osaka 564-8680, Japan, tobita@kansai-u.ac.jp
[3] Takashi Kiyota, Professor, The University of Tokyo, Institute of Industrial Science, Building B, 4-6-1, Komaba, Meguro, Tokyo 153-8505, Japan, kiyota@iis.u-tokyo.ac.jp
[4] Masataka Shiga, Assistant Professor, Nagaoka University of Technology, #1 Building of Machinery and Construction, 1603-1, Kami-Tamioka, Nagaoka, 940-2188 Niigata, Japan, shiga@vos.nagaokaut.ac.jp

Everyday drones excel in capturing real-time aerial imagery, presenting an invaluable asset for urgent landslide monitoring. Equipped with high-resolution cameras, these drones provide immediate visual feedback from disaster-stricken areas – Fig. 1. Rapid deployment of drones allows disaster response teams to obtain a comprehensive, bird's-eye view of the affected regions without jeopardizing human lives. The real-time streaming of imagery to a remote-control station facilitates prompt decision-making, enabling responders to strategize effectively and prioritize their efforts based on the dynamic conditions on the ground.

Figure 1. Illustration of the process of image collection and feature extraction for the creation of a 3D model for a bridge (Zekkos et al., 2018)

The purpose of the paper is to introduce three case-studies which focus on the implementation of drone survey (UAV) shortly after a seismic event – in this case, the 2023 Turkey-Syria earthquake. Two landslides (Islahiye and Tepehan) and one rockfall (Idili) have been included in the presented study – Fig. 2.

Figure 2. Map of the affected area (including fault and epicenter) – landslides and rockfall presented in this study

IOP Conf. Series: Materials Science and Engineering 1297 (2023) 012009 doi:10.1088/1757-899X/1297/1/012009

2. Real-time Aerial Imaging

Everyday drones excel in capturing real-time aerial imagery, presenting an invaluable asset for urgent landslide monitoring. Equipped with high-resolution cameras, these drones provide immediate visual feedback from disaster-stricken areas. Rapid deployment of drones allows disaster response teams to obtain a comprehensive, bird's-eye view of the affected regions without jeopardizing human lives. The real-time streaming of imagery to a remote control station facilitates prompt decision-making, enabling responders to strategize effectively and prioritize their efforts based on the dynamic conditions on the ground.

3. Multi-spectral and Thermal Imaging

Advancements in everyday drone technology have introduced additional capabilities, including multi-spectral and thermal imaging. By integrating specialized cameras, drones can capture data beyond the visible spectrum, offering essential insights for geotechnical monitoring. Multi-spectral imaging discerns variations in vegetation health and soil properties, providing vital indicators of slope stability. Conversely, thermal imaging can detect temperature anomalies, signifying potential subsurface issues, such as underground fires or unstable ground conditions, thereby enabling responders to identify critical areas requiring immediate attention.

4. Community Involvement and Empowered Response

The affordability and accessibility of everyday drones facilitate community involvement in post-earthquake disaster response. Local residents, volunteers, and concerned citizens can actively contribute to urgent landslide monitoring efforts by utilizing their personal drones to assist authorities in gathering crucial data. This collective response of drone-equipped civilians enhances the overall effectiveness of disaster management, complementing the efforts of professional responders and agencies. Moreover, the involvement of communities fosters a sense of ownership and solidarity during times of crisis, promoting a collaborative approach to safeguarding lives and vital infrastructure.

5. The 2023 Turkey-Syria Earthquake

On February 6th 2023 two major earthquakes struck southeast Turkey, M_w=7.7 Kahramanmaras and M_w=7.6 Elbistan, respectively. In these earthquakes, left lateral fault ruptures appeared along the East Anatolian Fault between Malatya and Hatay – distance between them is about 320 km – Fig 3.

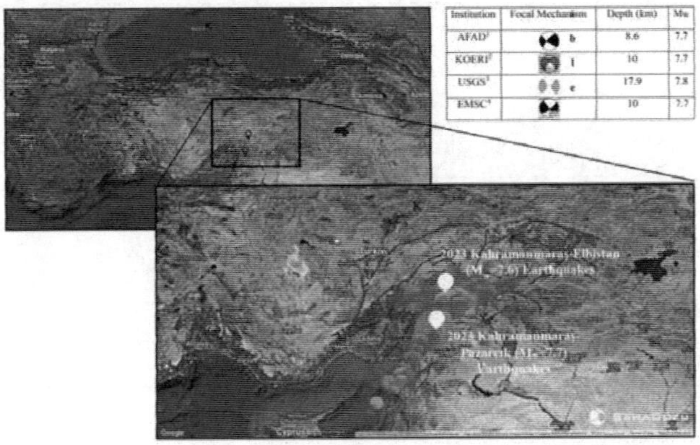

Figure 3. Map of Turkey. The epicenter of the February 6, 2023, Kahramanmaraş-Pazarcık, M_w=7.7, and Kahramanmaraş-Elbistan, M_w=7.6, (Çetin et al., 2023)

Unfortunately, due to the impact of these catastrophic events more than 50 000 casualties and 35 000 collapsed buildings have been reported since then. Moreover, the seismic motion affected an area of 99 000 km^2 on Turkish side where two million people were left without home. The PGA values which have been recorded by various stations show values as high as 1.2g as well as relatively spectacular maximum vertical component (PGV).

6. Islahiye Landslide

The landslide, which formed a natural dam due to the 2023 Turkey- Syria earthquake, has been observed at Degirmencik Creek Basin in Islahiye, where a bauxite mine on the top of the mountain is located – Fig. 4. The geological structure of the basin is primarily composed of Paleozoic-aged dolomites, Mesozoic-aged limestones, and Quaternary slope debris-cone, alluvial, and basalt units (Sinan, 2018). At higher elevation the formation is characterized by brown forest soils, while towards the graben floor, red Mediterranean soils and basaltic soils are observed. The altitude in the landslide region is around between 550 m to 1084 m above the sea level.

On a hillside approximately 3.5 km south-west of Islahiye, the slope on the left bank of the gorge has failed significantly to form a landslide dam. Digital elevation model and parameters obtained from drone (UAV) survey for this slope failure and landslide dam have been summarized on Table 1. The landslide occurred has a height of 220 m and a length of 480 m, with a maximum failure depth of 45 m. The failed slopes are shale and dolomite, and a number of rockfalls have been observed in the same valley as well.

The failed sediment deposit is shown on Figure 5. At the time of the March 31st survey, the dam crest has seemed to be compacted, with a buried drainage pipe upstream and a spillway constructed downstream. There has been no water in the spillway at the time of the survey, however, water flow has been observed downstream of the dam, which suggests that the blocked water is flowing downstream as seepage flow.

As a simple indicator of the stability of a landslide dam, (Ermini & Casagli, 2003) suggests a Dimensionless Blockage Index, *DBI*, as follows:

$$DBI = \log[A_{\times}(H/V)]$$

where A is the catchment area upstream the blockage (in km^2), H is the height of the landslide dam (in m), V is the volume of the landslide dam (in million m^3). Using the catchment area and the landslide dam dimensions, the *DBI* of the landslide dam is estimated to be 3.01, which is an uncertain level of stability according to the criteria given by (Ermini & Casagli, 2003). At the time of the authors' survey, the area of the blocked lake has not been large and the construction of the spillway has seemed to have maintained the temporary stability of the landslide dam, yet further monitoring is still considered necessary.

Table 1. Summary of dimensions of landslide dam near Islahiye

Area of landslide	0.103 km^2
Volume	996 000 m^3
Length	480 m
Height	220 m
Slope	24.6 ° (average)
Depth	45 m (maximum)
Height of Landslide dam	40 m

Figure 4. Overview of landslide dam at Degirmencik Creek Basin in Islahiye (37.007, 36.592)

Figure 5. Spillway constructed on the landslide dam

7. Tepehan Landslide

A landslide which has attracted interest among researchers in the recent months is the large failure in an olive grove near to Tepehan – Fig. 7. The region is mainly represented by sandstone, clayey limestone, claystone and marls (Kavuzlu, 2006) and the exact location of the failure zone is [36.161, 36.222]. It is considered as a block slide type of failure, most probably on a weak layer in a dip slope, which has allowed parts of the olive grove to remain intact despite large displacements – Fig. 6.

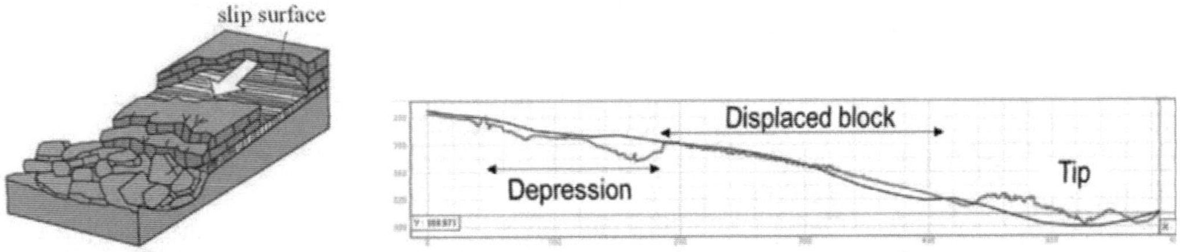

Figure 6. Failure mechanism of the Tepehan landslide and digital elevation model and ground surface before/after the landslide obtained by drone (UAV) survey

IOP Conf. Series: Materials Science and Engineering 1297 (2023) 012009 doi:10.1088/1757-899X/1297/1/012009

The weathered surface of the sandstones within Tepehan formation is light grayish-brown, while the freshly broken surface is yellowish-beige in color. The sandstones are moderately thick and layered. The unit is generally loosely cemented, with calcium carbonate ($CaCO_3$) acting as the cementing material. The sandstones contain gastropod and lamellibranch fossils with fragile shells that could be easily broken. The thickness of the Tepehan formation varies between 200 to 500 m and its age is identified as Middle Miocene, Langhian-Serravallian (Selçuk, 1985). It is believed that the units forming the Tepehan formation have been deposited in a shallow marine environment initially, which gradually deepened into a shallow-open marine environment.

Figure 7. Overview of the Tepehan landslide (36.161, 36.222)

Figure 8. Small landslide dam formed at the tip of landslide

The Tepehan landslide is approximately 500 m long and 100~150 m wide. The slope is composed of relatively brittle limestone, and fossil fish have been identified in the rock fragments. The landslide has also formed a small landslide dam at the tip, which is about 5 m higher than the original ground level due to the deposition of collapsed sediments.

The digital elevation model obtained from the drone (UAV) survey and the ground surface along Line 1 (Fig. 9) are shown in Figure 6. The hills surrounding are generally gentle and the slope of the landslide is approximately 4°. A depression zone at a depth of about 10~15 m has been observed up to about 100 m from the top scarp. It is also remarkable that the collapsed and moved soil block retains its original shape. The mechanism may be the presence of a weak thin layer at deeper levels, almost parallel to the ground surface, which has been weakened by the earthquake motion, causing the upper soil block to slide while keeping their shape.

Figure 9. Satellite image and Line 1 – used for the digital elevation model

The landslide in Tepehan is located on an extension of the fault that caused the earthquake, and the linear cracks in Figure 8 and Figure 9 not only match the fault direction, but also appear to be in a left lateral strike pattern, which is a characteristic feature of the fault. It is therefore possible that the occurrence of the landslide in Tepehan may have been influenced by fault movement, but this connection has not been clarified in the present study.

8. Idili Rockfall

The third case study shows a rockfall which is located near the village of Idili [36.991, 36.578]. Site visit has shown that rocks have blocked the road and it is temporarily closed due to this reason – Fig. 10. The drone (UAV) method indicates that when an earthquake occurs it is effective to fly a drone in some roads which are not easily reachable in order to verify the possibility for emergency escape in affected areas, especially in small towns and villages – Fig. 11.

Figure 10. Overview of the Idili rockfall (36.991, 36.578)

Further examinations have shown that the rocks present on the site are dolomites. Dolomite rocks are prominent sedimentary formations, predominantly consisting of the mineral dolomite ($CaMg(CO_3)_2$), distinguished by its calcium magnesium carbonate structure. Their genesis involves dolomitization, a complex process occurring in ancient marine environments where calcium-rich waters interact with magnesium-rich seawater or hypersaline lagoons, leading to the replacement of calcium ions in calcite with magnesium ions.

Figure 11. Application of the drone (UAV) survey for rapid detection of blocked roads by rockfall

The intriguing geological textures exhibited by dolomite rocks, comprising intricate patterns of crystals, nodules, and veins, have captured the attention of geologists, prompting in-depth investigations into their formation mechanisms. Beyond their aesthetic appeal, these rocks bear significant scientific importance as they serve as valuable archives of past environmental conditions. Their study offers insights into historical climate fluctuations and the evolution of ancient oceans over extended geological time scales.

Conclusion

The advent of everyday drones has ushered in a new era of democratized aerial assessment in urgent landslide monitoring following earthquakes. Their accessibility, user-friendliness, real-time imaging capabilities, geospatial functionalities, multi-spectral and thermal imaging, and community involvement have empowered diverse stakeholders to actively engage in disaster response. From first responders to civilians, the widespread adoption of everyday drones offers a powerful and inclusive approach to assessing damage, prioritizing rescue efforts, and enhancing the resilience of communities in the face of natural calamities.

As technology continues to evolve, the ongoing development and refinement of everyday drone capabilities hold promise for further expanding their role in disaster management. Ethical considerations surrounding data privacy, security, and airspace regulation will be integral to maximizing the potential of everyday drones while upholding societal values. Embracing this approach to aerial monitoring reinforces a unified response and fosters more resilient communities capable of confronting the challenges posed by post-earthquake landslides, exemplifying the positive impact of technology on disaster preparedness and response.

The presented paper suggests the use of the UAV method as an effective approach for rapid detection of landslides mechanisms and assessment of their geometry and dimensions. Its applicability has been demonstrated in three case studies – two landslides (Islahiye and Tepehan) and one rockfall (Idili), respectively.

References

[1] Çetin, K., Ilgaç, M., Can, G. and Çakır, E. (2023). *Preliminary Reconnaissance Report on February 6, 2023, Pazarcık M_w=7.7 and Elbistan M_w=7.6, Kahramanmaraş-Türkiye Earthquakes. REPORT NO: METU/EERC 2023-01*. Earthquake Engineering Research Center.

[2] Ermini, L. and Casagli, N. (2003). *Prediction of the behaviour of landslide dams using a geomorphological dimensionless index*. Earth Surface Processes and Landforms: The Journal of the British Geomorphological Research Group 28(1): 31-47.

[3] Kavuzlu, M. (2006). *Altinozu (Antakya) ve yakin civarinin tektono-stratigrafisi*. Master thesis, Cukurova Universitesi: 45-48.

[4] Selçuk, H. (1985). *Kizildag–Keldag–Hatay dolayinin jeolojisi ve jeodinamik evrimi*. M.T.A Rapor No. 7787, Ankara.

[5] Sinan, M. (2018). *Hassa-Islahiye arasinda gelisen birikinti konilerinin jeomorfolojik ozellikleri ve arazi kullanim durumu*. Master thesis, Mustafa Kemal University: 62-65.

[6] Zekkos, D., Greenwood, W., Lynch, J., Manousakis, J., Athanasopoulos-Zekkos, A., Clark, M., Cook, K. L., Saroglou, C. (2018). *Lessons Learned from the Application of UAV-Enabled Structure-From-Motion Photogrammetry in Geotechnical Engineering*, Vol. 4, Issue 4, p.254-274. doi: 10.4417/IJGCH-04-04-03.

IOP Conf. Series: Materials Science and Engineering 1297 (2023) 012010 doi:10.1088/1757-899X/1297/1/012010

Experiences and practices from application of rockfall remedial works at MKD national road network

Milorad Jovanovski, Igor Peševski, Jovan Br. Papić, Sead Abazi

University "Ss. Cyril and Methodius" in Skopje, blvd. Partizanski odredi No. 24, 1000 Skopje, MKD

jovanovski@gf.ukim.edu.mk

Abstract. Rockfalls, landslides and floods continue to have a growing share in the overall damages and losses, influencing considerably the national budget. Recent experiences and practices related to application of stabilisation measures and remediation of rockfall zones along MKD national road network are presented in the article. Analyses are based on case histories where the prognosis of geotechnical conditions is a key element in definition of remedial measures for rock slopes. Some issues related with Consequence Class, as suggested in Eurocode 7 202x, and the concepts of Acceptable Level of Risk are presented. Proposals in definition of Acceptable Level of Risk, through combining Probability of Failure, Reliability Index and Consequence Classes, are given in the article. These findings can help in development and/or improvement of rockfall, landslide and flood risk management strategies.

1. Introduction

Different types of mass movements like landslides and rockfalls, as types of natural hazards, frequently occur along road network. Together with floods in some road sections, they used to cause significant economic and life loss [1]. This is an important challenge worldwide, and a reason to develop several important research projects, with an aim to establish comprehensive methodology for solving these problems.

In MKD, special emphasis is given to the general design guidance for recovery of the most common types of damaged transport and water infrastructure contained in the country's first-ever Build Back Better manual [2]. Another important national document is "Guidelines for the Public Enterprise for State Roads" prepared by IMC Worldwide and supported by World Bank, which serves as technical assistance for climate resilience design. Its aim is to support the policy and decision makers in taking climate hazard resilience considerations for the purpose of road transport management and in the response plans [3].

Authors' experiences with rockfall hazards lead to conclusion that all risks must be minimised to acceptable level, using methodology for management in a systematic way. Different persons and stakeholders must be involved in the process [4], [5], [6].

Respecting the complexity of the problem, possible relations between Factor of Safety (*Fs*), Probability of Failure (*Pf*) and Reliability Index (β) with so called Acceptable Level of Risk (*ALR*) are presented below. The methodology is mainly intended for rockfalls, but some adaptations can make it applicable for other geotechnical problems, as well.

2. Research elaboration

The research methodology is based on analysis of typical rockfall events and applied remedial measures. The problem is specific because all measures involve some uncertainty in the input parameters, which depends upon large variation of geological conditions, heterogeneity, discontinuity properties, anisotropy, rock mass strength, stresses, groundwater conditions etc.

Among the other uncertainties that have to be considered in the design of protective measures in hard rock slopes are the loading conditions, as they affect the reliability of the analysis procedure. Moreover, sometimes problems need urgent measures which must be applied immediately after some event. The usual management approach applied in Public Enterprise for State Road (PESR) is presented in Fig.1.

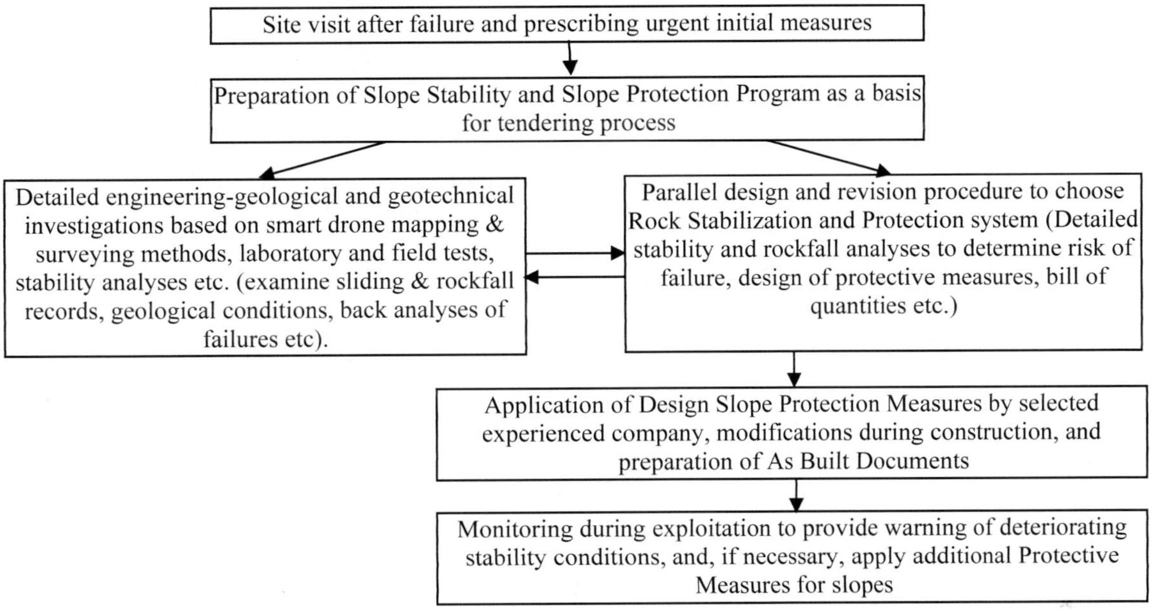

Figure 1: Flow chart illustrating the process of management of slope stability problems in PESR [4]

It is evident from Fig. 1 that immediately after a failure, a site visit is the first necessary step in order to prescribe urgent temporary (initial) measures. After that, typical phases of Slope Stability and Slope Protection Program preparation, investigation and design process that interfere one to each other and application of measures are being applied. Monitoring during exploitation is very important and necessary, because confirms the adequacy of the solution.

Methodology includes back-analysis of the past failures, prognosis failure modes, estimation of rock strength although joints etc. In brief, the methodology is based on application of several interrelated phases, using the following approaches:

- Analyses of possible kinematic modes of failure;
- Statistical analyses for defining probability distribution functions for all geotechnical input parameters;
- Defining Factor of safety;
- Defining Probability of failure and Reliability Index;
- Numerical analysis with Finite Element Method for some cases;
- Defining risk from sliding or rockfalls;
- Cost-benefit analyses for applying slope remedial measures;
- Definition of Acceptable Level of Risk.

The design procedures can follow different standards or recommendations, but the trend is, as much as possible, to use reliability analyses, suggested also in Eurocode 7 [8].

Preparation of Digital Elevation Model using smart technology that enables easy and productive mapping & surveying, mainly with drone techniques, is sometimes strongly recommended. These layouts serve then as a basis for detailed engineering-geological and structural-geological mapping, help in identifying potential unstable block positions, approximate volumes of blocks, and define kinematic conditions for failure. Slope stability analyses of discontinued media should be performed by using probabilistic approach, in order to see the level of slope's initial safety, obtaining values of *Fs* and *Pf.*

Beside *Pf*, the Reliability Index β is also commonly used as probabilistic measure of safety. If it is assumed that, after a probabilistic analysis, the *Fs* is normally distributed (i.e. it has a Normal Distribution), then the Equation 1 can be used to calculate β:

$$\beta=(\mu-1)/\sigma \qquad (1)$$

where: β – Reliability Index, μ – Mean *Fs* and σ – Standard Deviation of *Fs*.

It can be noticed that β represents the number of standard deviations that separate the mean *Fs* from the critical *Fs*. It can also be calculated using Log-normal Distribution.

3. Research elaboration

Rock Slope Protection Systems at critical sections at the country road network are composed of several main measures applied on certain slope areas, depending on the expected mode of failures.

The rehabilitation works are mostly consisted of construction of: rockfall barriers; shotcrete in combination with rock bolts and wire mesh; rock anchors combined with steel ropes; steel wire mesh and steel wire fences.

Particular set of circumstances, such as: rockfall volume, environment issues, influences on surroundings facilities, design loads etc., are considered in the analyses. The rockfall protection barriers are usually defined with risk analyses and adequate rockfall analysis.

One application for a regional road near the town of Veles in specific environment is presented in Fig. 2 and in Fig. 3. The application of measures is usually in difficult conditions, which asks for using special techniques of installation because of the pronounced morphology in most of the cases (Fig. 3).

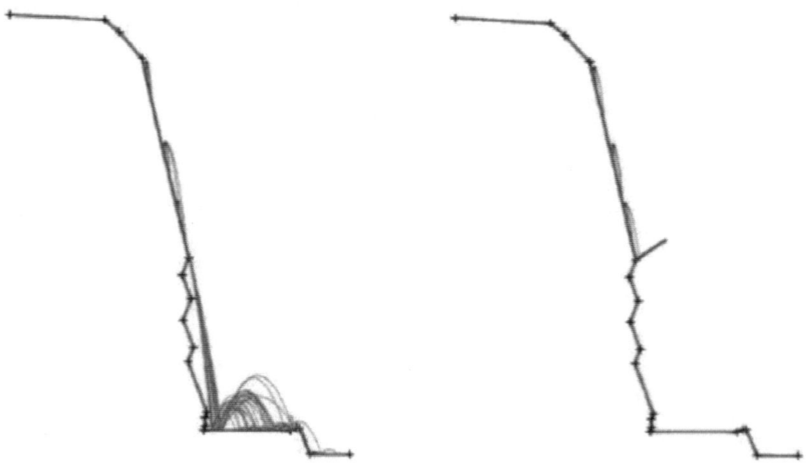

Figure 2: Example of 2D rockfall analyses using Rockfall software without (left) and with installed protective barrier (right)

Figure 3: Installed rockfall barrier at the upper part of critical slope in combination with wire mesh and rock-bolts at the lower part of the slope (left); Special technique of installation of the supporting elements (right)

Some specifics in remedial work applications are presented in Fig. 4 and Fig. 5. It can be noticed that the application of measures is usually quite difficult and complicated.

Here, an optimal "compromise" between acceptable level of safety and economics must be found, and maybe this is one of the main challenges from theoretical and practical points of view in road and rock engineering.

Authors' earlier experiences are summarised in several diagrams presenting relations between levels of *Fs*, *Pf* and definition of several classes of Risk Level (Fig. 6).

Figure 4: Application of protective measures at road from town of Strumica to town of Berovo, using specific combinations of wire mesh with rock bolt and steel ropes

IOP Conf. Series: Materials Science and Engineering 1297 (2023) 012010 doi:10.1088/1757-899X/1297/1/012010

Figure 5: Slope protection applied at the road to the arch dam Sv. Petka nearby Skopje by using technics of controlled blasting before application of shotcrete with wire mesh and rock bolts

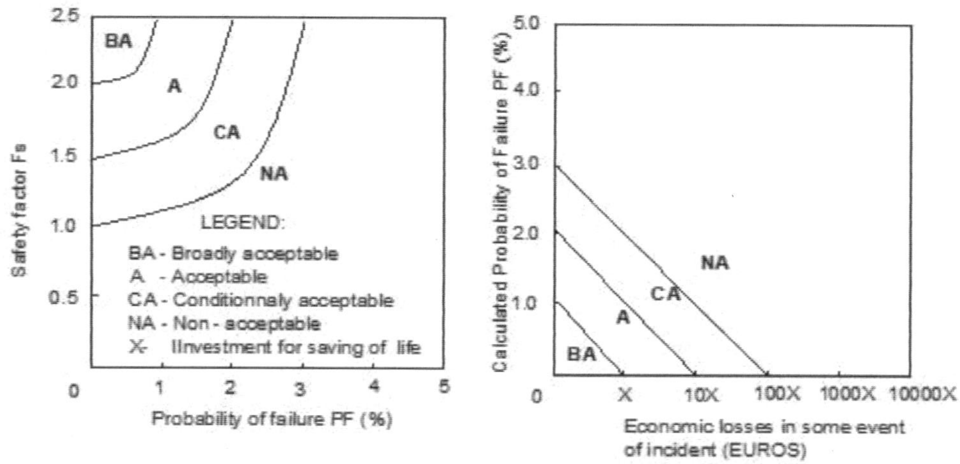

Figure 6: A concept for defining ALR: Mean value of Factor of Safety vs. Probability of Failure (left); Probability of Failure vs. Potential economic losses in some event of incidents (right)

The left portion of the diagram in Fig. 6 shows the mean *Fs* against *Pf*, while the right portion gives relation between *Pf* and potential economic losses for incident.

In both diagrams, there are zones marked as Broadly Acceptable (BA), Acceptable (A), Conditionally Acceptable (CA) and Non-Acceptable (NA) risk.

The zone with Non-Acceptable risk (NA) must be either avoided or reduced, irrespective of the costs.

Within the CA region, risks may be tolerated, as there are variants for solution optimization.

The value X in the right-hand diagram at Fig. 6 includes possible economic and life losses impacts and, from practice, it is connected with insurance to loss of life estimation. In fact, the term "value of life" shall be replaced with a term "investment to avoid life loses".

In Fig. 7, when it is needed to define acceptable (tolerable) level of risks, a simple way to define so-called Protection Effect (PE) is given with equation 2:

$$PE = (F_{rem} - F_{init})/F_{init}, \quad (2)$$

where: PE – effect of applied measures in increasing *Fs*; F_{rem} – *Fs* with applied remedial measures; F_{init} – *Fs* without applied measures (initial *Fs*).

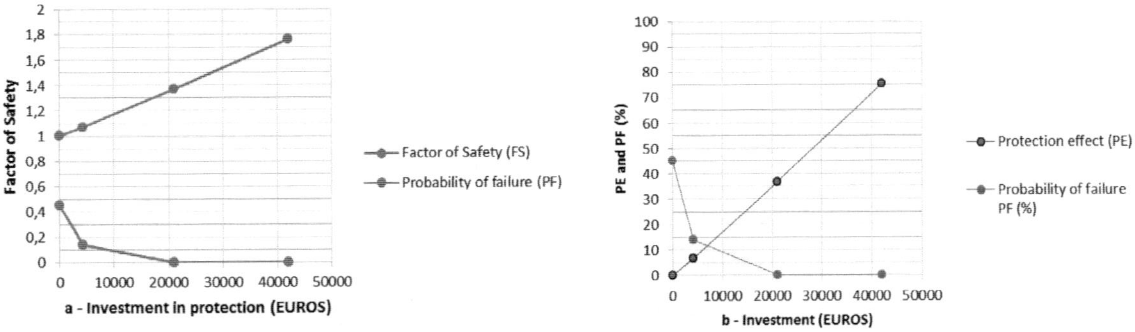

Figure 7. Influence of investment on *Fs* and *Pf*, for a case study analysed in [7]

Authors' suggestions for future works are related to including of reliability index, despite the fact that there is still no consensus regarding the suggested value for reliability index β and maximum probability of failure *Pf* for different types of geotechnical structures, as can be seen in [9] where some ranges are given. There are new trends that should be followed in the near future. For example, the next generation of Eurocode 7 incorporates the term Consequence Classes (CC) [10].

In Table 1, some ideas for future analyses are presented in a form of recommended value for β and *Pf* for hard rock slopes, where rockfall often occur. For that purpose, the data given in [9] was considered.

Table 1: Authors' recommendations for value of Reliability index β and Probability of failure *Pf* for hard rock slopes related to Consequence Classes in Eurocode 7 202x

Consequence Class	Reliability index (β)	Probability of failure (*Pf*) [%]
Lower CC1	1.0 - 2.3	$10 - 10^{-1}$
Normal CC2	2.3 – 3.0	$1 - 10^{-1}$
Higher CC3	3.0 – 5.0	$10^{-1} - 10^{-3}$

Summarised simple diagram is presented in Fig. 8.

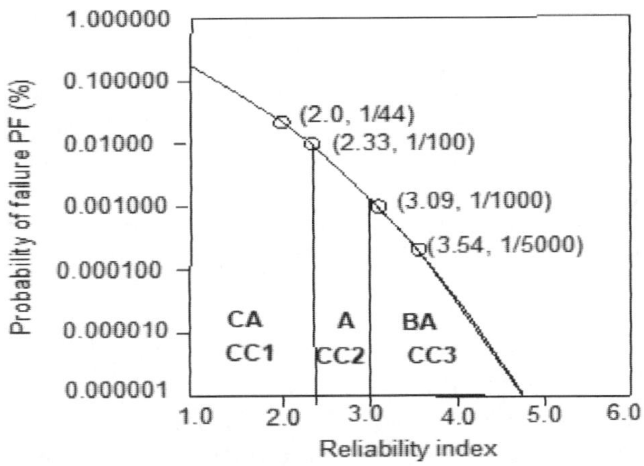

Figure 8. Diagram presenting zones of risk from Fig. 6 with Consequence Classes from Eurocode 7 202x, Reliability index and Probability of Failure values

One example to introduce a value for necessary protection effect is presented in Fig. 9.

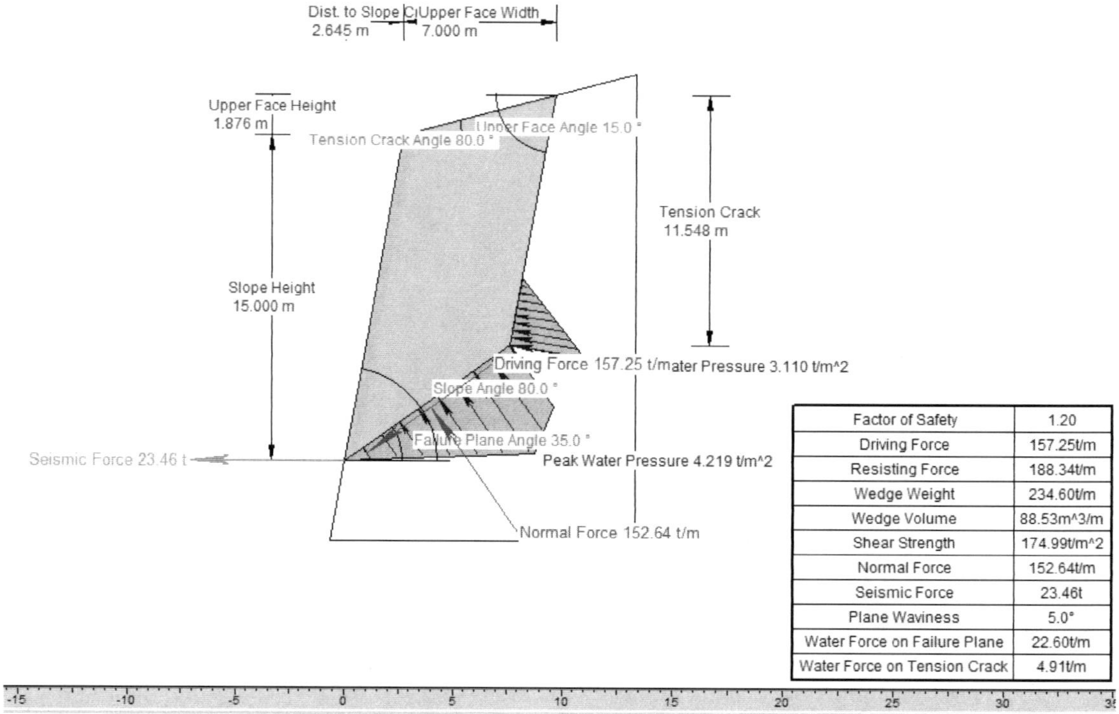

Figure 9. Case of probabilistic stability analyses for a slope at road to arch dam Sv. Petka: Probability of failure *Pf*=12%; Normal Reliability Index=1.165 which is in a zone CA in Fig.8

In order to reach Acceptable zone (A) from Fig. 8, and if accepted that the range of desired value for consequence class CC2 should be β_{des}=2.33 to 3, the necessary relative level of investment can be calculated. Because reliability index from Fig. 9 is $\beta_{init}=1.165$, the investment must be about 100% to 150% higher than for a case of initial values for slope construction (PE_{des}=(2.33–1.165)/1.165=100%).

4. Conclusion

The given analyses show that all experiences and practices can be useful in developing of new ideas, despite the fact that each case is unique and has to be considered in terms of particular set of circumstances. Beside upon detailed analyses, solutions should also be based on engineering judgement resulting from and guided by practical and theoretical studies of rockfall risks and stability of slopes. Authors believe that presented experiences and proposed approaches can help in establishing a more effective way for protection of slopes susceptible to rockfall instabilities.

References

[1] Peshevski, I.: Landslide susceptibility modelling using GIS technology, PhD Thesis, Faculty of Civil Engineering – Skopje 2015

[2] Popovska, C.; Jovanovski M.: Build Back Better Manual: Roadmap towards resilient transport and water infrastructure, EU Flood Recovery Programme, United Nation Developing Program 2018

[3] Abolmasov, B.; Pesevski, I.; White, J.R.; Panov, A.: Technical assistance preparation of climate resilience design Guidelines for the Public Enterprise for State Roads, Skopje 2019

[4] Nikolov, E.; Jovanovski, M.; Janevski, B.; Peshevski, I.; Velevski, A.; Milanovski, M.; Trajanovski, V.: An Approach for Rock Slope Protection on an area near City of Veles, Proceedings from 5th Symposium of Macedonian Association for Geotechnics, ISRM Specialised Conference, Ohrid, 2022

[5] Jovanovski, M.; Dimitrov, B.: Experiences, methodology and principles for rock slope stabilisation, case for a road to arch dam "Sveta Petka", Savremena Gradjevinarska praksa, Novi Sad 2010

[6] Jovanovski, M.; Peshevski I.; Papic J.; Abazi S.: An approach for slope protection on the access road to arch dam "Sveta Petka" in Republic of Macedonia, Naucni skup, Geoexpo, Sarajevo 2017

[7] Peshevski, I., Jovanovski, M., Nedelkovska, N., Lepitkova, S.: One approach in definition of acceptable level of risk for slopes in hard rocks, XVI Danube - European Conference on Geotechnical Engineering, 07-09 June, Skopje, 2018

[8] Vagnon, F; Bonetto, S.; Ferrero, A. M.; Harrison, J. P.; Umili, G.: Eurocode 7 and Rock Engineering Design: The Case of Rockfall Protection Barriers. Geosciences, 10(8), 305. 2020. doi:10.3390/geosciences10080305

[9] Fernandez, W.; Villalobos, S.; King, R.: Probabilistic Analysis of Slope Stability in Completely Decomposed Granite Residual Soils, Revista ingeniería de construcción, On-line version ISSN 0718-5073, Rev. ing. constr. vol.33 no.1 Santiago Apr. 2018, http://dx.doi.org/10.4067/S0718-50732018000100005

[10] Franzen, G; Garin, H.: Eurocode 7 – an updated framework to ensure reliable geotechnical solutions, 18th Nordic Geotechnical Meeting, IOP Conf. Series: Earth and Environmental Science 710 (2021) 012035 Publishing, doi:10.1088/1755-1315/710/1/012035

IOP Conf. Series: Materials Science and Engineering 1297 (2023) 012011 doi:10.1088/1757-899X/1297/1/012011

Depth profile of Eurocode LM1 Load and its influence on geotechnical structures

N Kerenchev[1], A Dobrinova[2]

[1] Assoc. Prof., Dipl.-Ing. PhD, Department of Geotechnics. University of architecture, civil engineering and geodesy – Sofia, Boul Hr.Smirnenski 1, Sofia 1000, kerenchev_fte@uacg.bg
[2] Cand. ing., University of architecture, civil engineering and geodesy – Sofia, Boul Hr.Smirnenski 1, Sofia 1000, albena2000@gmail.com

Abstract. This study concerns the impact of the moving loads LM1 on geotechnical structures like embankments and retaining structures related to cut and fill slope problem for road construction and maintenance. In many national standards or recommendations there is a depth dependant reduction factor for live road loads. Bulgarian road infrastructure agency and other institutions do not account for the load reduction with depth, which corresponds to higher cost of the infrastructure itself. This problem appears mainly when calculating the most common 2D tasks, while LM1 and the soil are 3D in nature. This paper presents two geotechnical structures an embankment and a retaining wall, where 3D FEM model is used. Using higher level soil model as hardening soil model with small strain stiffness, the study obtains the reduction of the tandem systems of LM1 and its influence on the slope surface or the structure itself.

1. Introduction

The characteristic value of LM 1 is a statistically extreme value [1]. In the calibration of the basic highway traffic loading the model LM 1 is the 1000-year return period value. It is also loaded with average dynamic loading factor with an approximate value of 1.4. The problem of missing real 3D to 2D loading reduction for traffic loads begins after applying the Eurocode standard more than 10 years ago. Well trained geotechnical engineer knows the significance of that reduction, but it could be different for different geometry conditions and should be assessed. Non geotechnical engineers and some Bulgarian agencies are reading the code inaccurate since not all geotechnical structures apply to bridge abutments and loading them according EC 1991-2 4.9.1 [2] with equivalent load q_{eq} of 60 to 90 kPa doesn't account for the reduction with depth. Some standards indicate other traffic load values when calculating slope stability or other embankment structures. The British Standard has old and new values respectively 10 and 20 kPa [3]. Finnish Transport Agency (2011b) applies that reduction factor according to fig. 1, where the value of the equivalent load q_{eq} is obtained based on the depth of the slip surface. After 7 meters the value of the combined LM1 (uniform distributed load UDL + tandem systems TS) is reduced to 12 kPa. Hence the dynamic loading factor and its damping with depth, and the geometric load distribution itself, the value of equivalent tandem system load of 10 kPa after 5m depth looks reasonable.

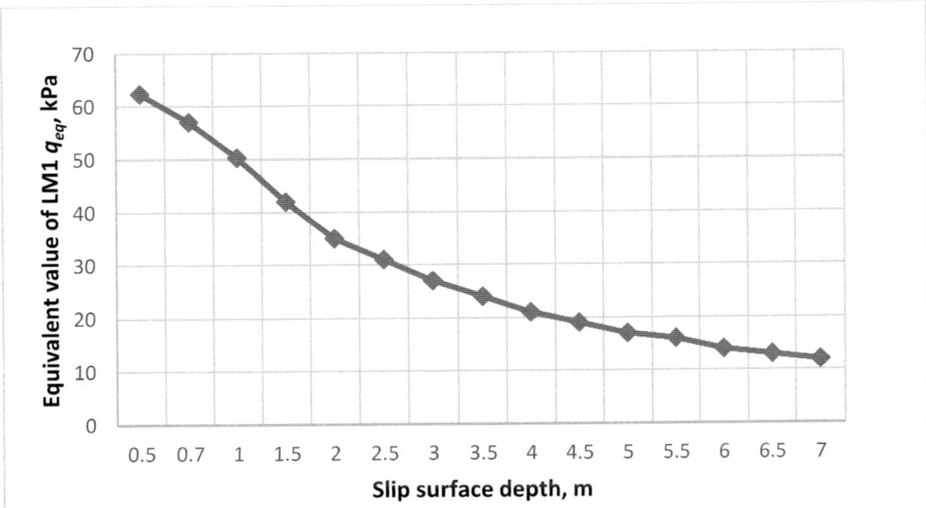

Figure 1. Uniform load at depth H, caused by traffic load diagram LM1, kPa – [4].

2. Models

For the purpose of finding the best reduction coefficient for the tandem system of LM1 with depth two different geotechnical structures are modelled. First structure could be described as regular embankment with height of 5 m placed on normally consolidated (NC) subsoil. The second structure corresponds to 5 m retaining cantilever wall with backfill on normally consolidated subsoil. Both structures represent the most common scenarios for geotechnical structures in road construction.

2.1. Geometry and loading

Figure 2 shows both model with their dimensions (enlarged). Note that the geometry of the models has extended boundaries (at least 20 meters apart from the TS loading points, not visible on the figure).

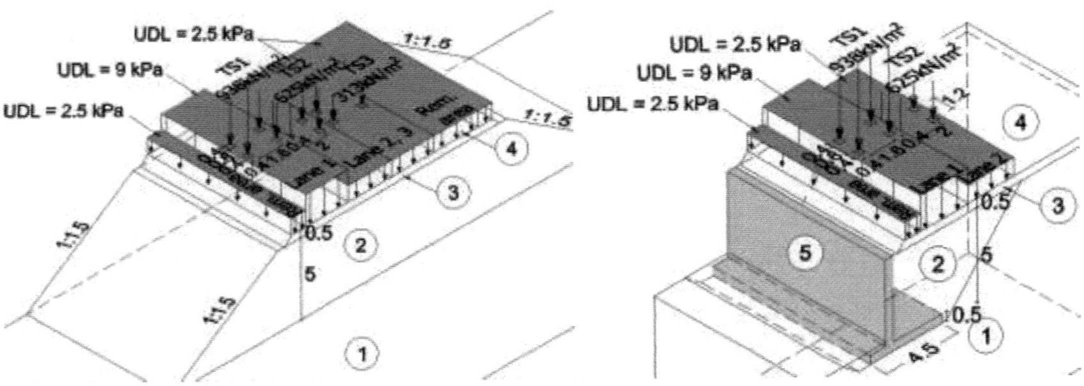

Figure 2. Embankment with height of 5 m and retaining cantilever wall with backfill both on normally consolidated subsoil.

The loading procedure is introduced into the model stagewise, since the software allows every loading condition to be separately applied. In the first stage the subsoil is present and the gravity loading of soil is applied. The second stage clears the displacements and places soil 2 or the cantilever wall respectively (the second model needs one more stage). The third stage applies soil 3 and asphalt concrete. Forth stage applies the uniform distributed load (UDL) and the fifth adds the tandem systems (TS).

IOP Conf. Series: Materials Science and Engineering 1297 (2023) 012011 doi:10.1088/1757-899X/1297/1/012011

2.2. Material properties

The materials match the most common types used for construction of road embankment and retaining structures. Soil types are: 1 – silty clay (NC); 2 gravely (drain) soil ; 3- well graded gravel (part of the road). Since modelling the non-ultimate stress-strain behaviour is mostly based on deformation criteria safety factors for the soil parameters are not applied. Table 1. presents the material properties for the soil, the asphalt concrete and the cantilever wall.

Table 1. *Material properties for Hardening soil model with small strain stiffness (HSS) for the soil and Linear Elastic material model for the asphalt concrete and concrete wall [5], [6] ,[7].*

Soil/Material			1	2	3	4. Asph. c. (*d*=0.20)	5. Wall (*d*=0.50)
Unit weight	γ_n	kN/m^3	18	18	18	23	24
Oedometer modulus	E_{oed}	kN/m^2	8000	40000	60000		
Elastic modulus	E_1	kN/m^2				500000	25000000
Modulus at 50% shear strength	E_{50}	kN/m^2	16000	40000	60000		
Unloading reloading modulus	E_{ur}	kN/m^2	64000	100000	120000		
Small strain shear modulus	G_0	kN/m^2	70000	120000	150000	208000	10420000
Poisson's ratio	v'	–				0.2	0.2
Poisson's ratio unl./rel.	v_{ur}	–	0.2	0.2	0.2		
Shear str. at 70% reduction of *G*	$\gamma_{0.7}$	–	0.00025	0.00015	0.00015		
Referent pressure	p_{ref}	kPa	100	100	100		
m parameter	m	–	0.7	0.5	0.4		
R_f parameter	R_f	–	0.9	0.9	0.9		
Over consolidated r.	*OCR*	–	1	1	1		
Earth pressure at rest $K_{0,nc} \approx 1-\sin\varphi$	$K_{0,nc}$	–	0.62	0.41	1.41		
Cohesion	c	kPa	30	1	5		
Friction angle	φ	°	22	36	36		
Soli group simbol			*siCl*	*saGr*	*Gr*		

Note that the asphalt concrete is an elastoplastic material, but since its tension pressure is not reached in the calculations, its properties in the table above remain elastic. The cantilever wall could also be modelled with non-constant geometry in height but it won't affect the results significant. Modelling the asphalt concrete and the cantilever wall is done using shell elements.

2.3. Meshing and other model conditions

Regularity of the geometry contributes to easy meshing process. The mesh is finer where the loads (TS) are applied and coarser at the boundaries. Near the structure itself some refinements were needed. The solution uses standard solver technique where no error refinements and correction were applied [8]. The hardening soil model with small strain stiffness contributes to more precise load spreading process and estimates the deformations and stress-strain relationship in the model in compliance with the realistic soil behaviour.

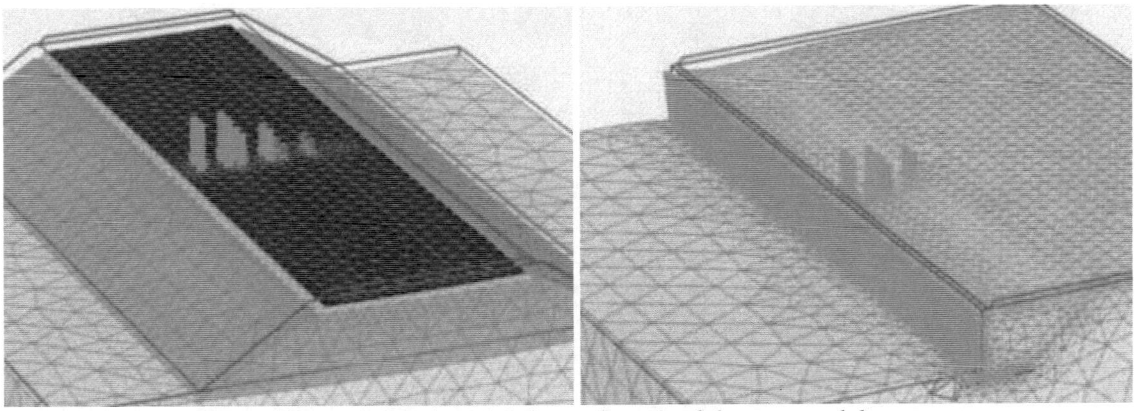

Figure 3. Mesh and deformed mesh of the two models.

3. Results

Based on the goal of the study, to find the depth load distribution factors of the tandem systems, the results are expressed mainly in form of stresses. The loading process described above is well suited to take into account the elasto-plastic and hardening/softening behaviour of the soil, as well as elastic distribution of the asphalt plate. Stresses generated by the previous (UDL) stage should be subtracted from the last full LM1 loading stage stresses. In result the stresses generated from the tandem systems are obtained.

3.1. Vertical stresses and depth dependent equivalent distributed load for tandem system

Depth dependant representative values of the tandem systems could be obtained by the vertical stresses for different depth profiles. Figure 4 shows the vertical stresses σ_{zz} equal to $q_{eq,TS}$ (equivalent distributed load from TS), with respect to depth, obtained from the embankment model located in the point y = 0 (the point between the TS loads).

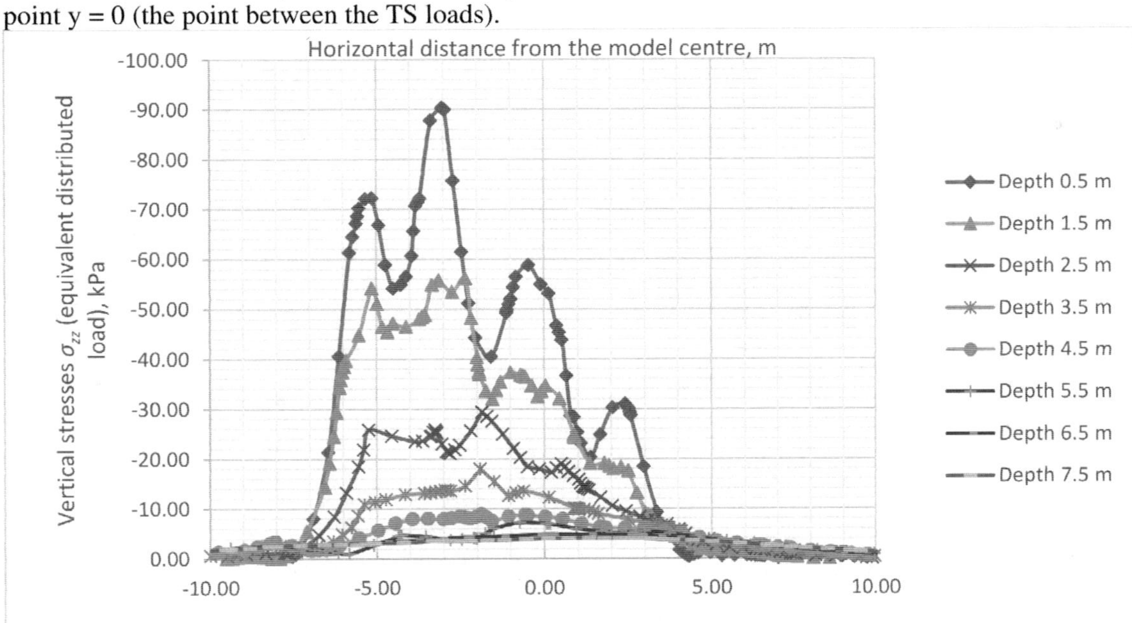

Figure 4. Vertical stresses $\sigma_{zz} = q_{eq,TS}$ (equivalent distributed load from TS), with respect to depth in kPa (embankment model)

Based on the graphical data the peak values of the tandem systems loads are significantly reduced and are not relevant after 2.5 meters of depth. The same depth corresponds to reduction of the tandem systems load to 30 kPa, where 0.5 m beneath the asphalt layers the value is still 90 kPa. Figure 5 shows the same type of diagram obtained from the cantilever wall model.

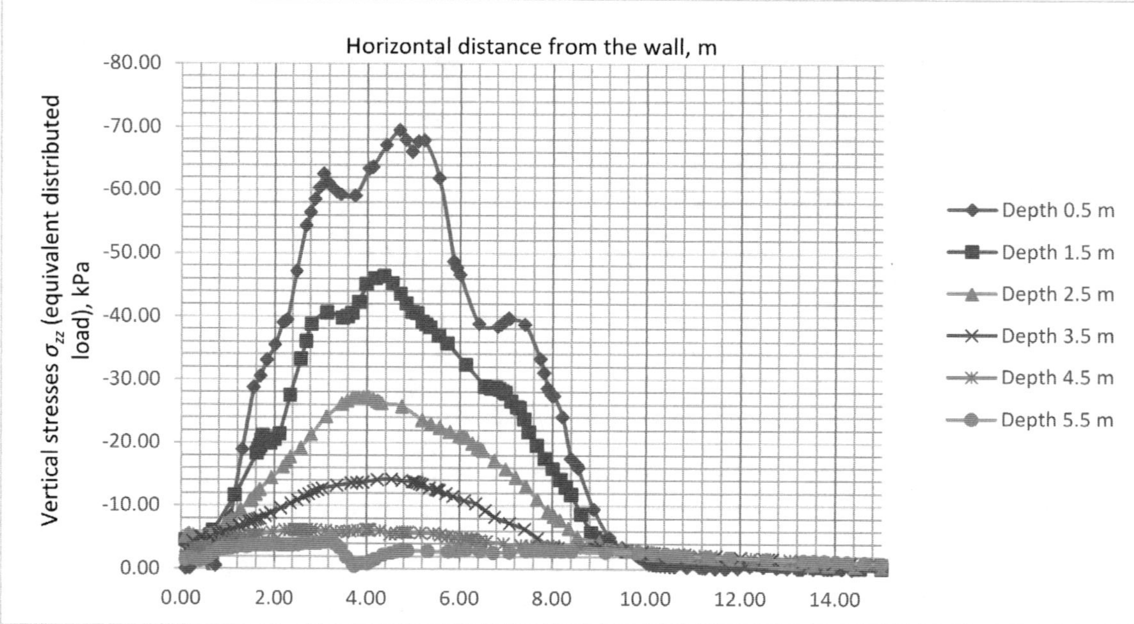

Figure 5. Vertical stresses $\sigma_{zz} = q_{eq,TS}$ (equivalent distributed load from TS), with respect to depth in kPa (cantilever wall model)

The cantilever wall model shows smoothed distribution, which could be explained with the wall influence and its redistribution. Figure 6 shows comparison of the vertical stresses $\sigma_{zz} = q_{eq,TS}$ for different y location for the embankment model. The diagram y = 0 corresponds to the point between the loads, while y = 0.6 is located exactly under the force point. Even at 0.5 m of soil depth the load shows higher σ_{zz} values under the midpoint (not under the loading point).

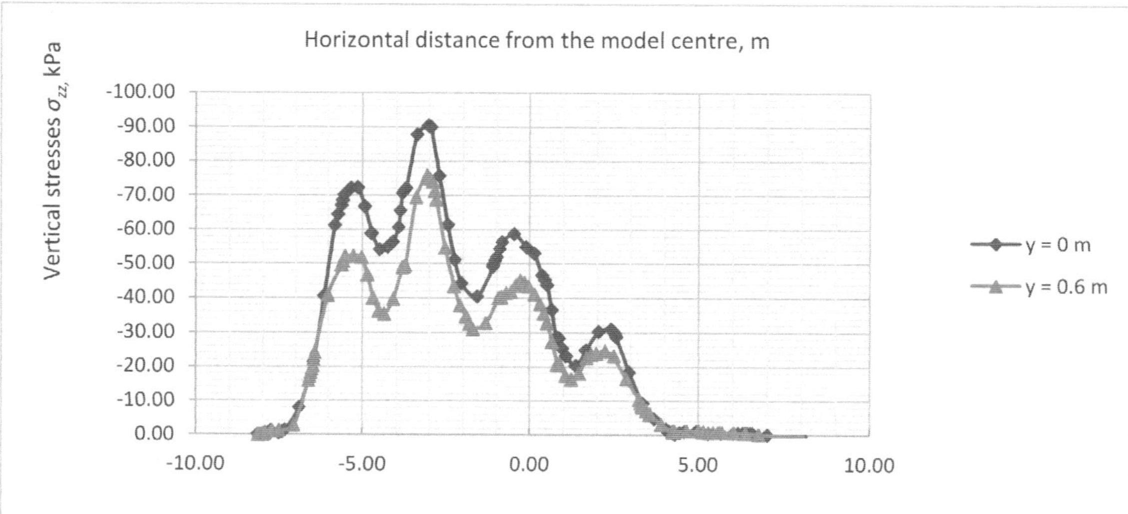

Figure 6. Comparison of vertical stresses $\sigma_{zz} = q_{eq,TS}$ (equivalent distributed load from TS) with respect to y position (embankment model), depth equal 0.5 m in kPa

Figure 7 shows a generalized graph of the maximal stresses σ_{zz} and the corresponding equivalent distributed load from TS $q_{eq,TS}$. The data shows heterogeneity for the first one meter and are more homogenous after two meters of depth. After that point the averaging line could be considered as relevant for the reduction factor estimation.

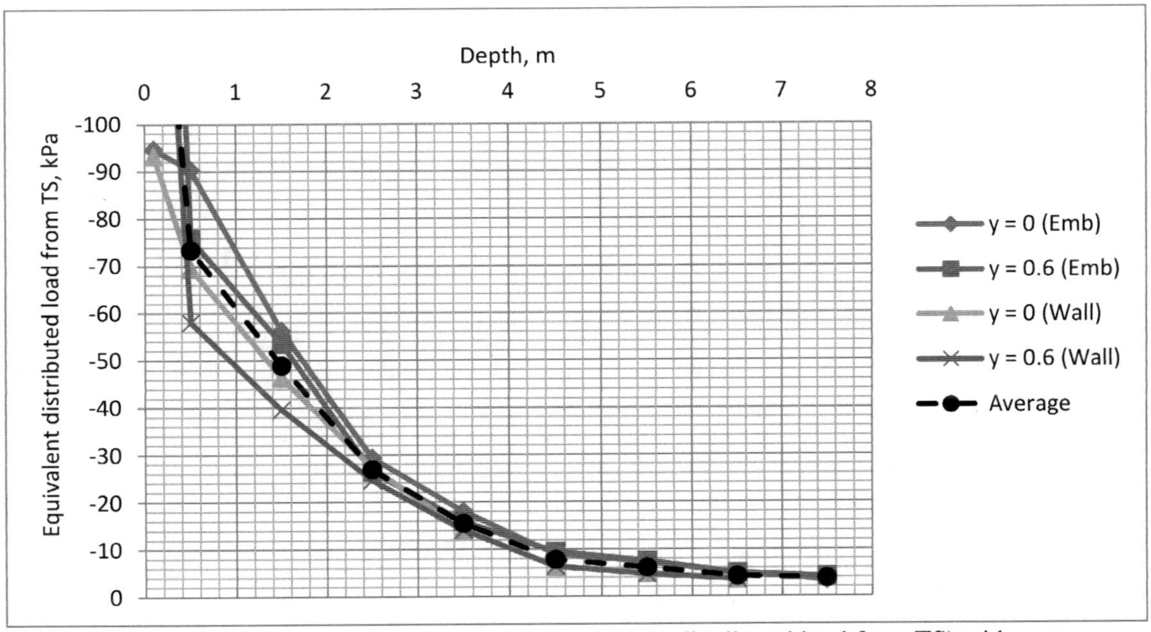

Figure 7. Maximal vertical stresses $\sigma_{zz} = q_{eq,TS}$ (equivalent distributed load from TS) with respect to y position from both models and averaging line in kPa

3.2. Horizontal effect on the retaining wall

The horizontal effect could be used when calculating earth pressure problems regarding cantilever structures. Figure 8 shows the distribution of the equivalent TS system horizontal stress with respect to wall height. With the current geometry (available road shoulder) and well working asphalt concrete pavement the active earth pressure from the tandem systems is less than 15 kPa and has more significant value around 10 kPa.

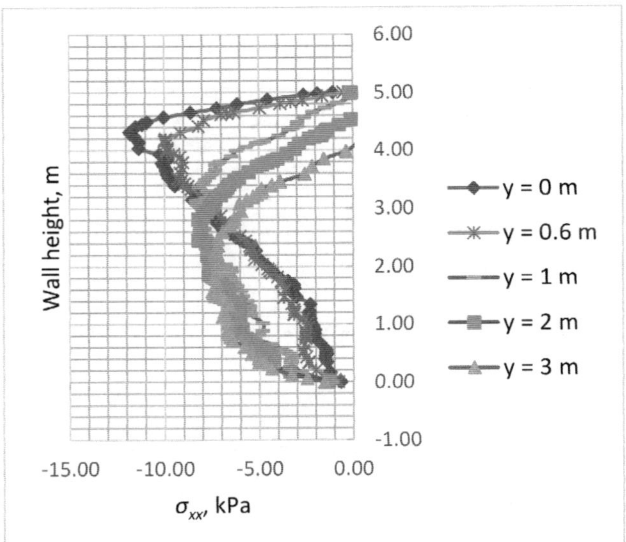

Figure 8. Horizontal stresses σ_{xx} representing active earth pressure from the tandem systems in kPa.

3.3. Bias and uncertainty

Based on the numerical nature of the study its uncertainty has to be taken into account. But since it is the only favourable method for estimation of the detailed load distribution in soil where the most sophisticated material models are used, this solution could be held as satisfactory. Different model conditions (mesh sizes, boundary condition etc.) and different soil characteristics, does not affect the results more than ± 10%, so a safety margin of 10% could be take into account.

4. Conclusions

This study show a significant difference compared to the direct (force divided by area) distribution of LM1 tandem systems in subsoil when calculating geotechnical road structures (as embankments and retaining walls). Most of the geotechnical engineers and some foreign agencies account for the reduction of the loads, but others [9] are waiting for the procedure to be written in their (ABC book) codes and do not allow the educated to use it.

In the current study:

- some standards and documents concerning tandem system depth reduction factor are reviewed;
- depth dependent equivalent distributed load for tandem system for an embankment and for a retaining wall is obtained;
- generalized depth dependent equivalent distributed load for tandem system is obtained;
- horizontal active earth pressure component of the tandem systems acting on the road retaining wall is obtained.
- general soil data for the HSS model of different soil types are presented;
- some minor points regarding the bias and the uncertainties are discussed.

The authors hope that this study could be taken into account for further development of the national annexes of Eurocode and other national norms.

5. References

[1] Bouassida Y, et. al. Bridge Design to Eurocodes, Worked examples. JRC Scientific and Technical Reports, 2010

[2] Eurocode 1, Eurocode 1991-2, 2003

[3] Shave J, Christie T, Denton S and Kidd A, “Development of traffic surcharge models for highway structure” Conference Paper, 2010

[4] K. Koivisto & J. Forsman, L. Korkiala-Tanttu, P. Salo & T. Perttula. Finnish national design guidelines fo georeinforcements following Eurocode – “Georeinforcement earth structures 2012”

[5] Setiawan A Suparma L B, Developing the elastic modulus measurement of asphalt concrete using the compressive strength test, AIP Conference Proceedings 1903, 050002 (2017)

[6] Kerenchev N, Markov I, “Determining the axial bearing capacity of pile based on common methods and comparison with pile load test”, Proceedings of the 3th international conference Vietgeo2016, 22-23 November 2016

[7] Ilieva K, Comparison of the deformations under a strip foundation with applying different constitutive models, Annual of the university of architecture civil engineering and geodesy, ISSN 1310-814X, 52, S1277-S1288, 2019

[8] Plaxis 3D, Material Model and Reference Manual, 2023

[9] Road Infrastructure Agency, Ministry of regional development and public works, Technical Specification, Sofia, 2014

Acknowledgments
The authors wish to acknowledge the conference organizing committee for the opportunity to present some significant results to the affected engineers and agencies. Acknowledgement also for the university institution that doesn't help but don't hinder such type of studies.

IOP Conf. Series: Materials Science and Engineering 1297 (2023) 012012 doi:10.1088/1757-899X/1297/1/012012

Innovative approach for surveying slope stability in road infrastructure rehabilitation

I Mitev[1], V Balev[2], V Penev[3], K Shtereva[4], L Moskovska[5] and K Kalchev[6]

[1]Assoc. prof. Department of underground construction, University of mining and geology "St. Ivan Rilski", 1700 Sofia, Bulgaria
[2] Chief assistant, Department of underground construction, University of mining and geology "St. Ivan Rilski", 1700 Sofia, Bulgaria
[3] Assistant, Department of underground construction, University of mining and geology "St. Ivan Rilski", 1700 Sofia, Bulgaria
[4] Assoc. prof. Department of surveying and geodesy, University of mining and geology "St. Ivan Rilski", 1700 Sofia, Bulgaria
[5] CEO, GeoMineData Ltd., 1320 Bankya, Bulgaria
[6] CEO, Alve consult Ltd, 1700 Sofia, Bulgaria

E-mail: [1]ivan_mitev@mgu.bg, [2]veselinbalev@mgu.bg, [3]Vladimir.penev@mgu.bg, [4]k_shtereva@mail.bg, [5]mining.softsolutions@gmail.com, [6]office@alveconsult.com

Abstract. The objectives of the repair, maintenance and rehabilitation of the road network in mountainous terrains focus on the study of the structural condition of the rock slopes. An innovative approach to structural mapping and analysis was implemented within the framework of a survey project for a road section of road II-35 Troyan-Karnare, known as "BEKLEMETO". This section begins with a long and steep descent from the Beklemeto saddle (at 1,525.5 m above sea level), passes through numerous bends on the southern rocky slope of the mountain and after 22.7 km descends into the northwestern part of the Karlovska valley, to the village of Karnare (at 562 m above sea level). The study of the structural disturbance provides an opportunity to present the groups of joints that have developed intensively on the rock slopes passing near the Republic Road II-35 in the section between Beklemeto and Karnare. Over the years, these slopes have been subjected to weathering, erosion and collapse processes, which have led to changes in the physical and mechanical properties of the rocks forming the slopes. The technology of mobile ground laser scanning, which is a suitable method for covering the complex characteristics of the terrain, was chosen for the research. The implementation of this technology was carried out through an integrated complex. Using this method, a total of 18 representative sections within the road II-35 "BEKLEMETO" were studied, and the number of measured cracks exceeded 13,000. The total length of the researched sections is over 10,770 m. In processing, short cracks with a length of less than 0.8 m, which do not affect the structural disturbance of the rock formations, are excluded. These data and studies are essential for the planning of effective measures to maintain and rehabilitate road infrastructure in complex mountain conditions.

1. Introduction

One of the key objectives in carrying out the repair, maintenance and rehabilitation of the Bulgarian national road network in mountainous areas is to explore the structural disturbance of the rock slopes.

In this scientific article, we focus on a certain section of road II-35 Troyan-Karnare, named ''BEKLEMETO''. This section passes through mountainous terrain with complex geology and requires special methods to study the rock formations along the road trajectory.

After analyzing the terrain which defines the section, we chose the technology of mobile terrestrial laser scanning and the Sirovision software for remote data mapping and analysis. The laser scanning technology uses an integrated complex that includes a RIEGL - VZ - 2000 hybrid laser system mounted onto a mobile platform (4WD vehicle). The equipment also includes two GPS receivers, a high-resolution digital camera "Canon" and a 3D inertial system 'figure 1(a) and (b)'.

(a) (b)

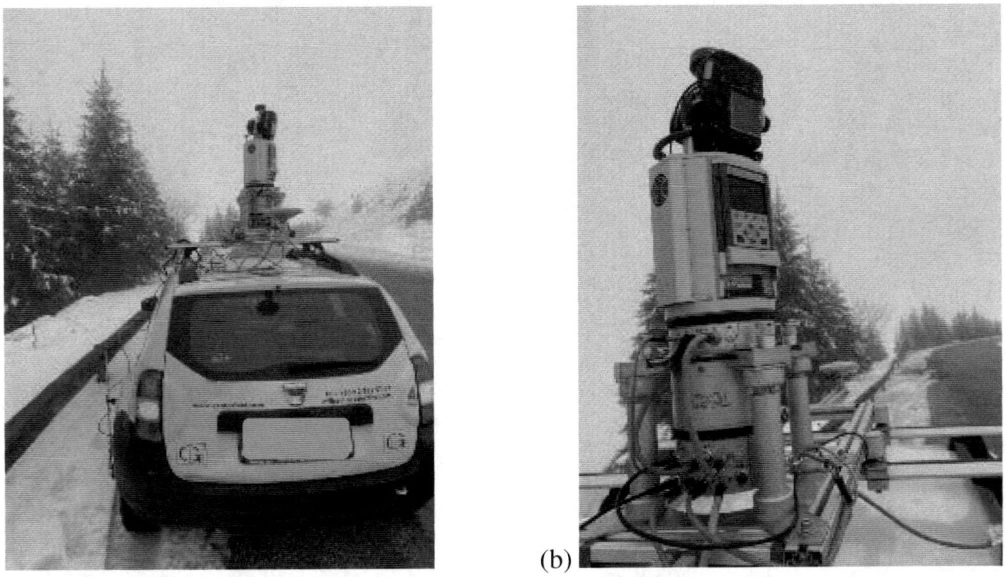

Figure 1. (a) Integrated laser scanning system "Riegl-VZ-2000" mounted onto a vehicle; (b) close view of the scanning system

For the more accurate determination of the trajectory of the complex, we used a GNSS receiver ("STONEX S900A") operating in static mode. For precise georeferencing (coordinate and height positioning) of the collected point cloud, we used pre-defined marks on the asphalt road, according to the accuracy requirements the project.

Due to the geometrical and geographical characteristics of the section, it was conditionally divided into two sections of equal length. The positioning of the ground control points is carried out by means of GNSS-measurements in RTK mode. Scanning of each of the sections, with some overlapping, is done twice in both directions, with the scanning device in a vertical position. In some cases where the GNSS signal was unavailable for sub-section scan because of dense tree vegetation, trajectory positioning was performed using the 3D inertial system. The total number of measured points in the cloud is 1,745,746,807 and they were mapped to different regions after filtering and processing. Pre-processing and filtering of information from the point clouds:

- Trajectory processing: The initial trajectory received by the IMU/GPS system is post-processed with specialized Post processing software, which presents it in real time with centimeter accuracy. It is georeferenced to a base station and aligned with individual laser records.
- Point cloud processing: The original point cloud 'figure 2 (a)' is filtered by excluding points outside the terrain such as vegetation, small objects, moving vehicles and others. The filtered result 'figure 2 (b)' is used to create digital terrain surfaces of the rock massif.

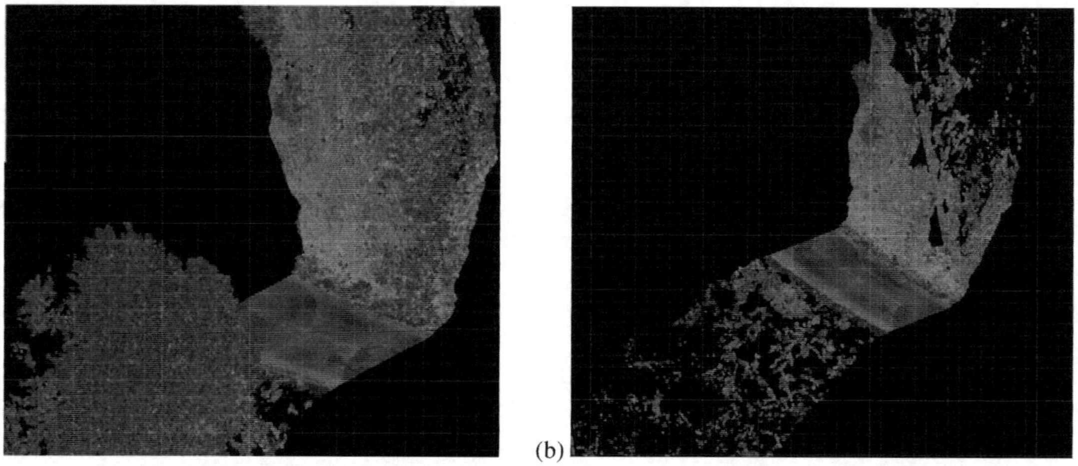

(a) (b)

Figure 2. (a) Original not filtered and (b) filtered point cloud, used for digital terrain surface creation

The Sirovision [1] software solution allows for remote mapping technique of the joint surfaces and digital record of all necessary structural data using just a desktop as well as kinematic analysis, mass and volume reporting.

This methodology provides important information for the study of the structural disturbance of the rock slopes and will provide a basis for planning of further maintenance and rehabilitation measures of the road section.

2. Study of the structural disturbance of the rock massif along the ''Beklemeto'' section

A study of the structural stability of the observed rock massif is of extreme importance due to the differences found between the mechanical properties of the mass measured in real conditions and those determined in the laboratory. This especially applies to parameters such as compressive and tensile strength, elasticity module and Poisson's ratio. In this context, it is necessary to understand the number and orientation of the loosening surfaces and their relation to classification assessments. In the presence of three or more joint systems, an elementary structural block is formed in the massif. The shape, size and orientation of the structural blocks affect the stability of the rock slopes, as these parameters are determinate in our study. In addition, naturally formed blocks affect the grade of fragmentation and the development of landslide and collapse processes.

To achieve the goals of the project, we carried out a survey and observation of the structural disturbance of the massif, focusing on each section separately. The field measurement of joints is carried out using a geological compass and tape measure. These measurements are carried out on a section of the rock escarpment at a pre-selected site called the "Measurement Station". The Measurement Station is marked with a thin rope or line forming a "measuring line". This line is duplicated with a tape measure, to facilitate reading the distances from the start of the station to the corresponding joint. During the measurement, the joints that cross the measuring line and have a length greater than 0.5 m are recorded. The indicators that characterize the spatial position of the joints are determined with the geological compass, which measures the parameters: dip angle (DIP=β), dip direction (DIP DIRECTION) and propagation direction (STRIKE=α). The desktop work consisted of joints surface mapping, establishing of joint system sets, conducting of kinematic analysis, mass and volume calculation and reporting of kinematically free structural features using the software solution Sirovision 'figure 3' in a combination with Dips v8.0 [2].

Figure 3. Desktop mapping of joint surfaces and structural data record using Sirovision

Mass measurement of joints has three main objectives: to establish the number of joint systems, their orientations, and to assess the joints' condition. Joint systems in rock mechanics are a group of repeating joints with apparently parallel or near-parallel directions and dip angles. Joints with different orientations that do not fit into any established system are called random. The condition of the contact surfaces of the joints determines the grade of mass deformability, its strength and the sliding possibilities. The number of joint systems and their orientations determine the structural blocks.

3. Laser scan data processing

Data processing and presentation of the study results are carried out using the software products Dips v8.0 [2] and Sirovision [1]. The joint data is plotted on an equal-area polar stereonet, where the joint plane is presented as a point called pole on a lower hemisphere. The lower hemisphere projection is a primitive circle/sphere outline graph divided into radial sectors through 10° and having North-South and East-West Reference Lines, where the inclined planes are displayed as great circles and show the strike direction of the joint. Nine concentric circles through 10° indicate the dip angle of the joint plane, which increases from the center to the outer circle meaning that a horizontal plane dipping 0° will be plotted as a great circle on the primitive circle of the net and a vertical plane dipping 90° will be plotted as a straight line passing through the center of the plot. For better understanding and interpretation of big data is most suitable the usage of poles-to-planes technique which we applied to this study. Instead of plotting the great circle of a planar structure, we plot its pole which represents the projection of the right-angle vector to the respective plane. The equal-area polar stereonet is the most suitable for representing the poles of joints because it does not require rotation or overlapping. Clusters of joint poles are well distinguished on the stereonet and this is the first step in obtaining the average orientation of individual joint systems. We used the Schmidt method to group and contour the poles of the joints, determine their density and group them into joint sets. The scatter range is determined with the Terzaghi correction, which is most often 15°. Contouring involves superimposing a square grid on top of the equal-area net. The smallest circle in the center of the equal-area net, representing 1% of its total area, is placed at the points of intersection of the grid lines. The number of poles that fall within this circle is counted for each intersection. Thus, the outlines of the different density groups in up to 6 contour intervals are obtained 'figure 4(a)'. The other option for presenting the results of mass joint measurement is the Rose diagram 'figure 4(b)'. This graph is a circle labeled from 0 to 360° divided by sector radii through 10°. The measured orientation data (strike only) is grouped to the nearest 10-degree sector, and their number is represented by the size of the "rose petals" mirrored on either side of the center of the circle. This method allows representation of the strike only, without including the dip angle. In our study we used two different software solutions to better represent, analyze and interpret the collected data [3].

IOP Conf. Series: Materials Science and Engineering 1297 (2023) 012012 doi:10.1088/1757-899X/1297/1/012012

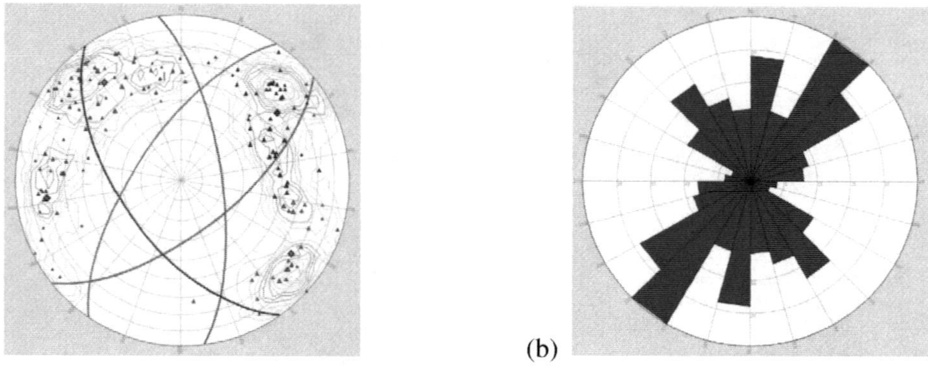

(a) (b)

Figure 4. Structural data representation of the observed area (a) on a Stereoplot and (b) on a Rose plot [2]

The combined use of Dips v8.0, Sirovision and data presentation methods allowed us to study in detail the structural disturbance of the rock massif and determine important parameters that are crucial for engineering decisions after assessing the stability of the rock slopes and ensuring the safety of the road section.

Sirovision is a Slope Stability Analysis tool that enables engineers and geoscientists to identify and quantify those areas on the rock slope where unplanned failures and movements can occur. This software enables geotechnical engineers to automatically detect all types of unstable features, such as wedges, blocks and toppling hazards on a rock face based on the structures that have been mapped. Sirovision enables the input of domain boundaries with individual geotechnical characteristics (density, cohesion, friction), pore pressure grids which is an important parameter in the rock mass assessment. An important feature of Sirovision is the ability to work with complex structures that can be non-linear. This allows for a more realistic representation of the rock geology and structures in the massif.

Mapped structural data is used to detect and visualise wedges and blocks which daylight on the rock face. Geotechnical data from domains, pore pressure grids and complex structures are used to compute a Factor of safety for each detected feature and provide a stability classification. Volume, mass and sliding vectors for each wedge and block are also an output. The value of using Sirovision is the ability of the engineering staff to identify the physical size, location and stability classification of structural features in the rock slopes (i.e., it will be clear where failures are most likely, as well as the volume of materials involved).

This information is valuable for understanding the scale of potential hazards and for geotechnical engineers to design and maximise ground support efficiency with confidence that the efforts are focussed in the correct areas and that safety and risk are paramount considerations. It is also important to road design engineers to design cut widths and slope angles to maximise the safety aspects and risks associated with the slope stability.

In summary, Sirovision is a powerful tool that provides geotechnical engineers and geologists with fast, reliable and accurate data for rock slope stability analysis and risk management of rock mass failure hazards.

4. Field study

The field "in situ" assessments of Joint Roughness Coefficient (JRC) are performed by comparing the joint surface with standard profiles proposed by Barton [4], [5]. This comparison is made visually by selecting the profile that most closely matches the profile of the natural disturbance 'figure 5'. For the longest joints that exceed the scale of the proposed profiles, the JRC value is determined along the entire length of the joint 'figure 6' [6].

IOP Conf. Series: Materials Science and Engineering 1297 (2023) 012012 doi:10.1088/1757-899X/1297/1/012012

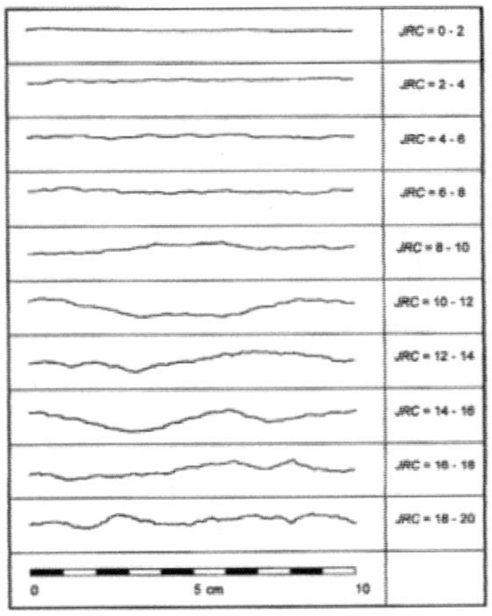

Figure 5. Relationship between the JRC and the determined profile

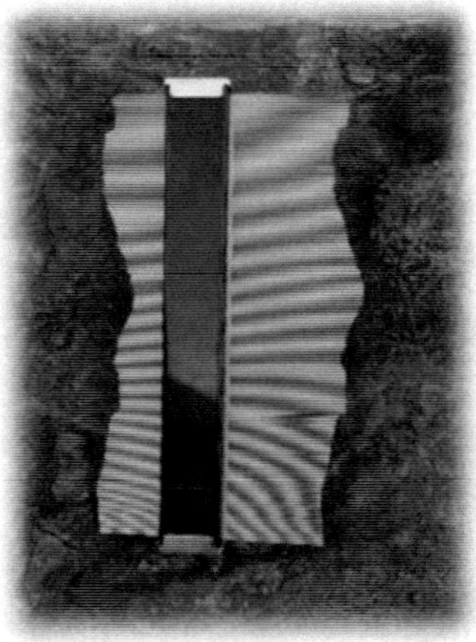

Figure 6. An alternative method for determining JRC depending on the amplitude of the roughness and the length of the joint.

The Joint Compression Strength (JCS) is a coefficient that represents the normal compressive strength of the joint wall. For the determination of JCS, recommended methods according to ISRM (The International Society for Rock Mechanics and Rock Engineering) include the use of a Schmidt hammer 'figure 7'. A Schmidt hammer is a tool used to determine the uniaxial compressive strength of rock [7], [8].

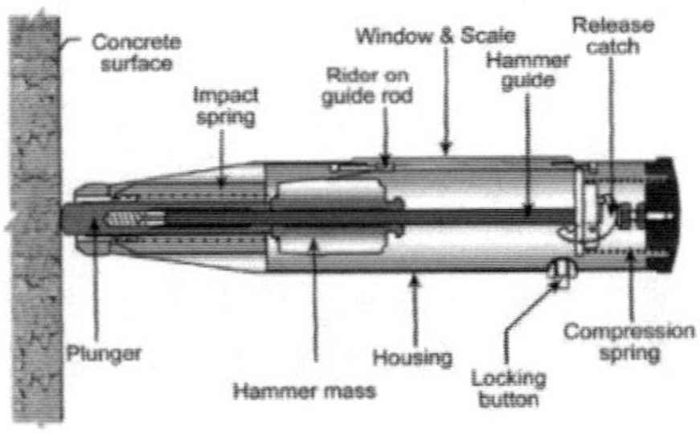

Figure 7. An example of the Schmidt rebound-hammer

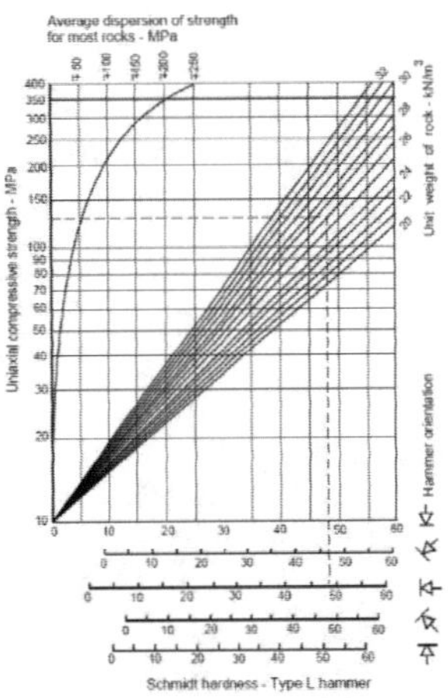

Figure 8. JCS estimation chart from the Schmidt hammer readings

Schmidt hammer results (L) are used to determine uniaxial compressive strength as a function of rebound. For reading and interpretation of the results is applicable the estimation chart showing the correlation between Schmidt hammer rebound number, hammer orientation, UCS and Rock density 'figure 8' [9].

These JRC and JCS field assessments are important for the analysis of rock slope stability and for the preparation of geotechnical evaluations in engineering projects that involve rock formations [10].

IOP Conf. Series: Materials Science and Engineering 1297 (2023) 012012 doi:10.1088/1757-899X/1297/1/012012

5. Results

The study of the structural disturbance of the observed massif was carried out in three meanings, aiming the determination of faulting grade of the slopes and establishing the number and orientation of the joint systems.

The first line of research was related to the establishment of the global structural faulting of the area. The entire area was properly observed, being divided into sections for more detailed study. In some of the rock slopes, where it was not possible to map the variety of joint surfaces due to the presence of safety nets, erosion or erosional deposits, random joint measurements were carried out in places with safe access.

The second line of research aims to check whether there is a spatial difference in the grade and nature of faulting between different sections. A comparison is made for the joint sets in each section.

The third line of research is related to the qualification of the general structural disturbance section by section.

The following parameters were determined in these studies:

- Strength of the joint walls using a Schmidt hammer, which is used to measure the resistance of the rock surface to normal stress.
- Geological Strength Index (GSI), which is a numerical parameter used to evaluate rock geomechanical properties and structural jointing.
- Joint Roughness Coefficient (JRC), which is used to estimate the roughness of the joint surface.

Joint Compression Strength (JCS) or also known as the normal compressive strength of the joint wall is a parameter used to obtain the uniaxial strength of the rocks forming the rock slope in the considered section. These studies and measurements are of key importance in determining the structural failure of rock slopes and in conducting geotechnical assessments, which are necessary in the planning and deployment of engineering projects and safety measures in rock massifs.

Applying some existing technologies as a combination of a novel integrated system in this project was a key factor for our study to obtain quick, safe and accurate results. We combined the point cloud collected with a laser scanner to create a real 3D surface of the observed rock slopes, import the surfaces into Sirovision to map the joint surfaces, record the dip and dip direction parameters and automatically analyze and report the slope stability parameters. This technique saved us more than 200% of the time needed to a field mapping and measuring, and protected the field geologists of rock failure hazards during field work.

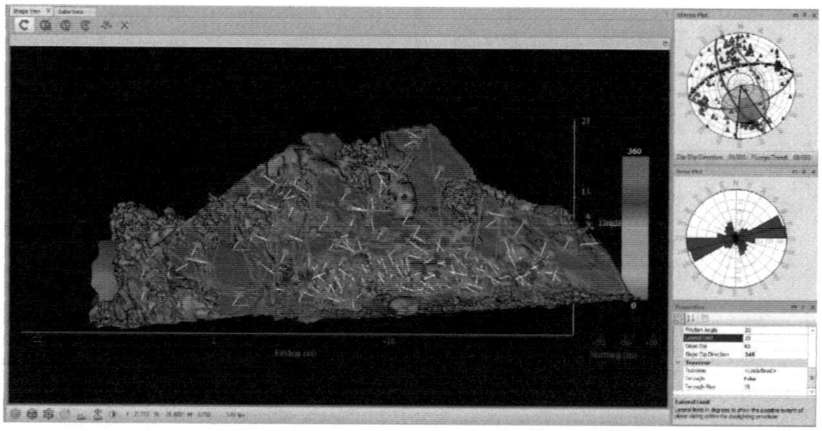

Figure 9. A digital 3D model of a rock slope representing desktop mapped structures and Orientation Sets classification

The joint surfaces are visible on the digital rock slope 3D model, structures are easily mapped and classified into Orientation Sets based on their dip and dip direction, here shown in red, blue, green and cyan 'figure 9'. On the right-hand side of this image are displayed the respective stereonet plot and the rose plot of all surfaces mapped in Sirovision for the selected sector.

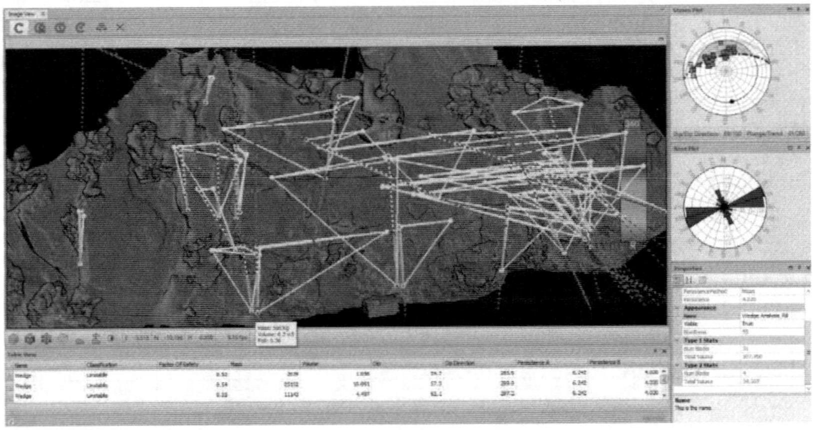

Figure 10. A 3D model of the same rock slope representing detected wedges formed between the mapped structures.

Based on the mapped joint surfaces and structural data the Slope Stability tool (kinematic analysis function) of Sirovision has detected wedges formed between the structures 'figure 10'. Structural wedges and blocks are locally unstable areas classified by type, based on a cirtain factor of safety. The classification of wedges takes into account various parameters such as the presence or absence of friction between the surfaces of the joints and the grade of stability. The possible wedge types are:

- Stable with friction: These are wedges that are stable and do not pose a high risk of falling because the friction between the surfaces of the wedges is high enough to keep them stable.
- Stable without friction: These types of wedges are also stable, but there is no friction between their surfaces.
- Unstable with friction: These are wedges that are unstable and pose a high risk of falling. The presence of friction is not enough to keep them in a stable condition.
- Unstable without friction: These wedges are also unstable and pose a serious risk of falling, because of the total lack of friction between the adjoining surfaces.

The other benefit of using this specialized software was our ability to calculate and report the mass and volume of potential hazardous rock masses recognized as wedges or structural blocks.

In addition, the visualization of wedges, their classification and volume reporting are of utmost importance for risk assessment and the design of appropriate engineering measures to ensure the stability and safety of rock slopes. Road design engineers furthermore are able to effectively use this information to make informed decisions and take the necessary actions to minimize the risk of falling unstable rock masses.

6. Conclusions

The main conclusions and recommendations for the observed area based on the selected approaches and methods are the following:

Cutting and construction of rock slopes along roads are specialized tasks that require expertise and specialized planning. This is necessary to ensure the safety of people traveling on the road as well as facilities located near the slopes. The collapse of loose rock blocks from rock escarpments can cause serious hazards to the movement of vehicles and people. It is therefore necessary to take appropriate measures to prevent collapses and manage their consequences. A variety of rock slope repair and maintenance techniques can be used to prevent potential failure hazards. The size of the structural blocks of rock escarpments can vary significantly from section to section. The design of reinforcement systems should consider the minimum and maximum size of the structural blocks for each section. Rock slopes are classified into two categories - Category I with a low risk of collapse processes and Category II with a high risk of collapse processes. For Category I slopes, measures should be taken such as pre-collapsing the loose rock blocks and placing safety nets and/or "pocket walls" in the specific sections. For Category II slopes, further research and analysis is required to determine appropriate ground support measures. When designing and sizing fence nets and reinforcement systems, it is important to take into account the specific characteristics of the rock slopes in each of the sections. Potentially endangered areas above the rock slopes are observed in some of the studied sections. These areas should be specifically marked and taken into account when planning signaling and safety measures.

In summary, the provision of rock slopes requires a complex approach, taking into account the specific conditions of the area and the characteristics of the rock massifs. Research and analysis played a key role in determining the necessary measures to strengthen and ensure the safety of travel on "Road II-35 "BEKLEMETO"" and the environment.

The presented information and sources emphasize the importance of laser scanning in combination with appropriate specialized software as a modern and powerful method for mapping and analyzing rock samples and rock slopes. Laser scanning provides detailed and accurate 3D models of the rock surface and joints, which is extremely useful for geotechnical engineers in assessing the stability of rock samples. The determined structural characteristics such as joint roughness coefficient (JRC), joint strength coefficient (JCS) and structural block sizes are essential for the analysis of rock mass behavior. Laser scanning allows precise measurement and documentation of these parameters, which contributes to the greater accuracy of geotechnical assessments. The laser scanning-based approach allows fast and safe observation of a large volume of rock surfaces, which is of particular importance in research in inaccessible and dangerous areas. This helps to reduce risks to personnel as well as planning of more effectively appropriate rock slope stabilization and safety measures. Also, laser scanning integrated with systems such as Sirovision creates opportunities for automated detection and analysis of structural features and unstable elements such as wedges and blocks that pose potential hazards to road infrastructure and people.

Contributing all its information and capabilities, laser scanning is a powerful tool that helps improve the precision and effectiveness of geotechnical assessments for rock slope stability. This method has the potential to significantly improve the design and implementation of engineering solutions that affect ground support and ameliorate problems.

Reference:

[1] Datamine 2015 Sirovision - Improving Safety and Minimising Risk in Mining
[2] Dips v.8.0. 2023 Rocscience, Canada
[3] RSData 2023 Rocscience, Canada
[4] Barton N and Choubey E 1977 The Shear Strength of Rock Jointed Model in Theory and Practice *Rock Mech.* **2**
[5] Barton N 1976 The shear strength of rock and rock joints *Int. J. Rock Mech. Min. Sci. & Geomech. Abstr.* **13** 1-24
[6] Patton F D 1966 Multiple Modes of Shear Failure in Rock Proc. *1st congr. Int. Soc. Rock Mech.* ***1*** Lisbon 509-13
[7] Bandis S, Limsden A and Barton N 1982 Experimental Study of Scale Effects on the Shear Behaviour of Rock Joints *Int. J. Rock Mech. Perg. Ltd.,* London
[8] Barton N 1982 *Shear Strength Investigation for Surface Mining*, London
[9] Deere D U and Miller R P (1966). *Engineering classification and index properties for intact rock.* (Illinois: University of Illinois Urbana) Ai*r Force Weapons Laboratory Technical Report*

[10] Barton N, Bandis S 1990 Review of Predictive Capabilities of JRC-JCS Model in Engineering Practice *Rock joints, proc. int. symp. on rock joints*, Loen, Norway, (eds N. Barton and O. Stephansson) 603-10 Rotterdam: Balkema

Evaluating the Long-Term Plastic Strain Accumulation in Reclaimed Asphalt Pavement (RAP) Application for Road Subbase Construction: A Cyclic Triaxial Loading Study

A Głuchowski[1], K Zając[1], W Sas[1]

[1] Water Center, Warsaw University of Life Sciences – SGGW, 166, Nowoursynowska, 02787 Warsaw , Poland

andrzej_gluchowski@sggw.edu.pl, wojciech_sas@sggw.edu.pl

Abstract. Reclaimed Asphalt Pavement (RAP) has gained significant attention in recent years as an environmentally sustainable solution for road construction. However, the main concerns are regarding its long-term performance and the potential impact on the road subbase. This study aims to investigate the behavior of RAP under cyclic triaxial loading conditions, specifically focusing on the accumulation of plastic strain over time. The research methodology involved laboratory testing using a cyclic triaxial apparatus to simulate the stress conditions experienced by RAP in road subbase applications in reference to AASHTO code. Several RAP specimens with varying degrees of compaction were subjected to repeated cyclic loading while monitoring the associated plastic strain accumulation. To provide a comparative analysis, conventional subbase materials were also tested under the same conditions. The findings indicate that RAP exhibits a notable disadvantage in terms of long-term plastic strain accumulation when compared to conventional subbase materials. The cyclic triaxial loading tests revealed a substantial increase in plastic strain accumulation over extended periods, suggesting the potential for long-term deformations and compromised performance of the road structure. This accumulation of plastic strain could lead to increased maintenance requirements and decreased service life of the road. Based on the results obtained, it is recommended that careful consideration be given to the utilization of RAP in road subbase construction in areas subjected to heavy cyclic loading conditions. Additional research and development efforts are necessary to explore potential mitigation strategies and optimize the use of RAP, ensuring sustainable and resilient road infrastructure.

1. Introduction

Road design requires extensive studies concerning road materials' properties. Each road layer must provide enough support to carry the traffic loading into the road subgrade. The road layer constituting the bottom part of road construction generally is a soil or/and aggregate mix which must present high stiffness properties. A cyclic triaxial test is employed to evaluate material stiffness design parameters for the soil material. The cyclic triaxial test (CTRX) can simulate stress and strain conditions that result from traffic load on soils and bituminous mixtures. For measuring stiffness properties during one loading cycle, the resilient modulus is calculated (1). The resilient modulus M_r is a quotient of cyclic deviator stress Δq to cyclic resilient strain observed in one cycle ε_r Eq. (1):

$$M_r = \frac{\Delta q}{\varepsilon_r} \tag{1}$$

The resilient modulus and permanent deformations are important properties for pavement design and can be determined through cyclic triaxial test. Exact M_r determination makes possible the designer to prognose the structural behaviour (2). The most reliable method for M_r testing of whole road construction is Long Term Pavement Test Protocol (LTTP) P46. The laboratory methods are less accurate and enables to test the stiffness properties only for one material. The laboratory method bases on empirical correlations like CBR test (3,4) or cyclic CBR test (5). The CTRX test is most reliable test for M_r estimation, but the initial loading conditions and test procedure often causes differences. The CTRX test requires to establish the following test procedures: (i) sample preparation, (ii) saturation conditions (iii) drainage conditions, (iv) confining pressure values (v) cyclic loading amplitude and frequency, (vi) cycle number, (vii) post-cyclic shearing resistance test. Often the CTRX tests are conducted with different protocols requiring prior conditions determination (6). Among resilient modulus which represents the elastic part of material properties, the plastic part is represented by deformation characteristics. The plastic deformation is observed in practice as road rutting which greatly contributes to serviceability time reduction. Asphalt layers in flexible road constructions has a viscoelastic characteristics and plastic part of deformation has a viscoplastic characteristic (7). Road rutting prediction relies on rutting prediction models which are a function of number of cycles *N*. The most common model is Barksdale semi-log model (8) which uses primary stage of rutting and two empirical constants Eq. (2):

$$\varepsilon_p = a_1 + b_1 \log(N) \tag{2}$$

Where, a_1 and b_1 are empirical constants which are regression coefficients.

The road building industry consumes large amount of materials. Therefore, natural aggregate replacement is highly beneficial. Construction and Demolition (C&D) materials like crushed concrete have considerable increase in popularity as it is reused either in concrete production or as unbound road subbase. In the case of asphalt, according to the information provided by the European Asphalt Pavement Association (EAPA), approximately 47% of the produced Reclaimed Asphalt Pavement (RAP) is reused in new asphalt concrete. The high level of RAP recycling is since reclaimed asphalt concrete has the same or sometimes even better properties than asphalt concrete without the addition of RAP (9). RAP is a granulated composite geomaterial acquired due to the replaced deteriorated roads (10,11). Usually, this material type is stored on landfills because the RAP reusing is limited to about 30% of RAP waste stream. Higher RAP application in asphaltic layers requires design of supplementary additives (9,12,13).

Physical properties are affected by: (i) old bitumen presence, (ii) bitumen content, (iii) aggregate gradation, (iv) type of RAP acquiring/crushing, (v) temperature, (vi) RAP grain shape. These characteristics are impacting such properties as: optimal moisture content, Los Angeles abrasion, specific gravity, dry unit weight. RAP physical properties can differ greatly from its natural counterparts and therefore are extensively studied. Nevertheless, for road construction application as a subbase, RAP material must be studied in respect to stiffness and strength properties. The main issue concerns the material performance during long-term cyclic loading and for that matter, the CTRX test on RAP with various physical properties. For instance, literature reports that adding RAP to unbound aggregates can be beneficial in improving the strength parameters of such mixtures. Adding RAP to the mixture affects the value of the bulk density of the granular skeleton and its natural moisture content, and it also results in a decrease in the CBR (California Bearing Ratio) value (14).

Asphalt rubble is most obtained through surface milling, excavation or are crushed from existing asphalt pavements or returned from job sites. Milling produces a loose material with grains smaller than 50mm (14). Crushing yields block-like material, which requires further processing, including crushing. The main difference between milling and crushing, apart from technological aspects, is generally the greater number of fine fractions in the case of milling (15). Milling, produces grains that may not be fully covered with asphalt or are only partially covered. When crushing asphalt blocks, the resulting aggregate is usually fully covered with asphalt (16). An inherent issue with milling and crushing is the change in grain size, as the grains themselves undergo fragmentation during these processes. As a result, RAP is evenly graded and has a slightly higher content of fine fractions.

In this article we investigate the RAP response to CTRX test, focusing on the accumulation of plastic strain over time. The goal of the research is to find the long-term RAP plastic strain characteristic and to evaluate the possibility to use the plastic strain accumulation models.

2. Materials and Methods

The research material in this study was sieved and then mixed in proportions that corresponds with the requirements for unbound mixtures according to both national and international standards. The resulted mixture of RAP fractions had a sand and gravel gradation (grSa). The grain size distribution curve is presented in **Figure 1**. Coefficient of curvature C_C was equal to 0.4 and coefficient of uniformity $C_U = 25$.

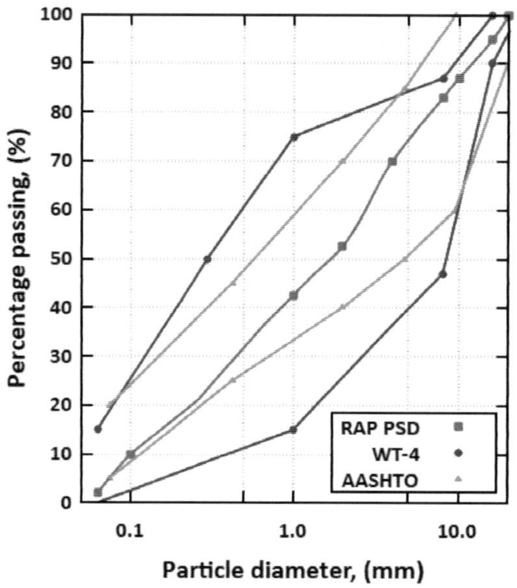

Figure 1. Particle soil distribution for RAP tested in this study (solid red line, material was first sieved into fractions, and then the universal soil grain composition was composed), green line-American standards requirement (17), blue line Polish standards requirement (18).

RAP material was obtained from an asphalt road surface crushed from rubbles thorn off from a section of an asphaltic road. The physical properties laboratory evaluation included the determination of compaction characteristics using the Proctor method, and the specific density. The physical properties are presented in **Table 1**.

Table 1. Physical RAP properties in this study.

	RAP
Optimal moisture content OMC, (%)	9.2
Maximum dry density ρ_{dMAX}, (g/cm^3)	1.85
Specific gravity G_S, (-)	2.65

2.1. Cyclic triaxial test (CTRX)

The CTRX tests were performed under a series of consecutive loads according to the AASHTO T-307 guidelines (19), applying the test sequences presented in **Table 2**. The testing procedure involved conducting tests at a minor principal stress $\sigma_3 = 20.7$, 35 and 69 kPa and at three increasing levels of deviator stress q within the minor principal stress range. The cyclic loading is performed at a frequency of 1Hz and for 100 cycles. The tests were carried out for the soil under optimum moisture conditions.

The samples were in unsaturated state. The soil specimen was first subjected to the minor principal stress by applying the desired pressure in the apparatus chamber (σ_3 = 103.4 kPa) maintained for an hour, and then the procedure presented in **Table 2** was followed.

Table 2. Cyclic loading procedure adopted in this study, based on AASHTO T-307 (19).

Stage number	Minor normal stress σ_3, (kPa)	Max deviator stress q, (kPa)	Number of cycles N, (-)	Frequency f, (Hz)
1	20.7	20.7		
2	20.7	41.4		
3	20.7	62.1		
4	35.0	20.7		
5	35.0	41.4	100	1
6	35.0	62.1		
7	69.0	20.7		
8	69.0	41.4		
9	69.0	62.1		

3. Results and discussion

The results of CTRX test are presented as a relationship between accumulated plastic strain ε_p and number of cycles depicted on **Fig. 2** and resilient modulus M_r and number of cycles presented on **Fig. 3**. The ε_p-N relationship shows that the plastic strain accumulation is stress dependent what confirms, (i) higher strain accumulation in stages 7 to 9, (ii) higher strain accumulation at stages 3, 6 and 9. The first effect is caused by confining pressure σ_3 the second effect is caused by higher deviator stress q. Nevertheless, in the case of the first effect at lower confining stress level (stages 1 to 6) is less obvious and seems that at σ_3 = 69kPa (stages 7 to 9) the forces between RAP grains could overcome the asphalt coating strength and would be a reason for roughly 5 times bigger plastic strain accumulation.

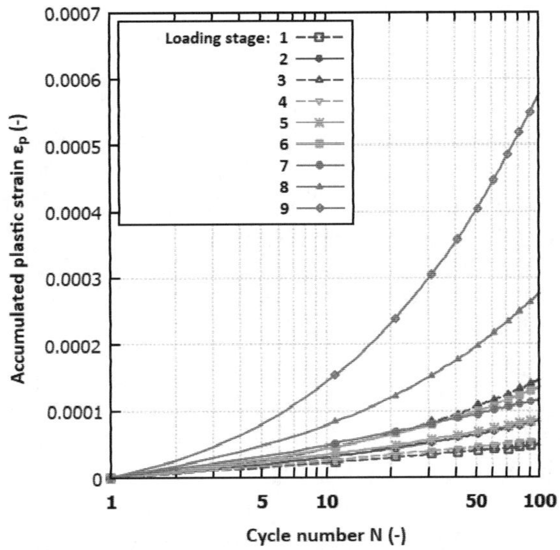

Figure 2. Plastic strain accumulation during CTRX tests for RAP material at three different confining pressures (i) σ_3 = 20.7kPa (blue), (ii) σ_3 = 35kPa (green), (iii) σ_3 = 69kPa (red).

3.1. Plastic strain accumulation

The lowest plastic strain accumulation was observed for lowest deviator stress value in all three confining pressure levels tested in this study. The M_r value extends the findings from accumulated plastic strain analysis (see **Fig. 3**). In the case of tests at σ_3 = 20.7kPa, the highest M_r value observed and therefore, the highest stiffness observed was for stage 1 where q = 20.7kPa and is the lowest for adopted test program. For tests in higher confining pressure levels the reversed trend is noted. The highest stiffness is observed for the tests at the highest q equal to 62.1kPa. The overall M_r value is determined by initial RAP stiffness, for subsequent cycles, the M_r value decreases in logarithmic manner and the magnitude of stiffness degradation differs among different confining pressure levels. To characterise the M_r degradation, the Degradation modulus δ was adopted which is defined as a quotient of resilient modulus in N-th cycle ($M_{r,N}$) to resilient modulus in first cycle ($M_{r,1}$) and is presented at **Fig. 3**. The akin degradation modulus characteristic has been observed in the case of CTRX tests at σ_3 = 35kPa.

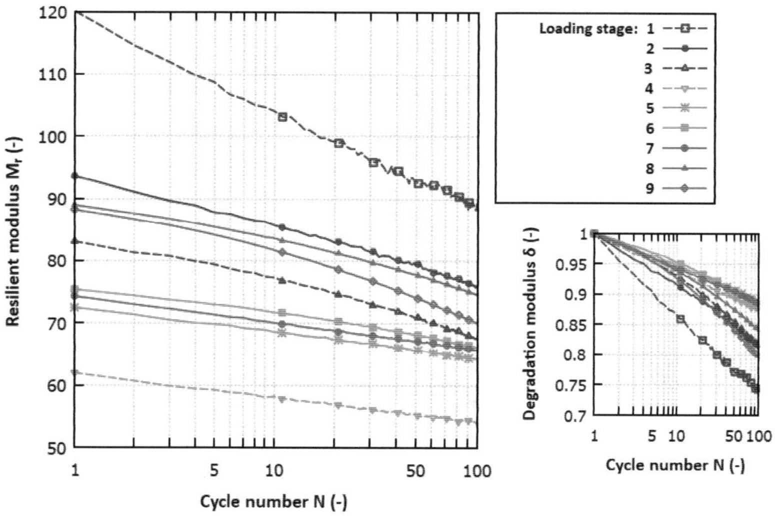

Figure 3. Resilient modulus and degradation modulus characteristics - CTRX tests for RAP material at three different confining pressures (i) σ_3 = 20.7kPa (blue), (ii) σ_3 = 35kPa (green), (iii) σ_3 = 69kPa (red).

Rapid stiffness decrease was spotted for stages 1-3 at lowest confining pressure. The high stiffness tends to be reduced more than the sample with lower initial stiffness like in the case of stages 4 to 6.
For prognosis plastic strain accumulation, we conducted a correlation analysis which led us to adopt power family function Eq. (3) as a model for ε_p calculation in subsequent cycles:

$$\varepsilon_p = a \cdot b^{\frac{1}{N}} \cdot N^c \qquad (3)$$

Where a, b and c are empirical constants. Detailed value of empirical parameters and standard deviation along with R^2 value presents **Table 3**. The power function curve was chosen because of the high correlation of all CTRX. The high R^2 and low standard deviation makes able to prognose the accumulated plastic strain in further cycles. On **Fig. 4** we are presenting the results of prognosis for tested samples. Adopted model was used for ε_p prediction after $1 \cdot 10^8$ cycles what may give the information about the fatigue RAP performance. For loading conditions at stages 1, 2, 4, 5, 7 after long-term repeating loading, the accumulated plastic strain is not grater than 2% what may indicate, that RAP would perform relatively well when deviator stress would not exceed 20.7kPa at all the confining stress levels. Additionally, deviator stress level at q = 41.4kPa would be acceptable for confining pressure σ_3 = 20.7 and 35kPa what seems to be counter-intuitive since higher confining stress would be expected to have positive effect on RAP resistance to cyclic loading. This finding confirms damaging σ_3 effect to RAP grains.

IOP Conf. Series: Materials Science and Engineering 1297 (2023) 012013 doi:10.1088/1757-899X/1297/1/012013

Table 3. Empirical parameters for accumulated plastic strain calculation for RAP under cyclic loading.

Stage no	a	b	c	St. dev., *s*	R^2
1	0.000016	0.101	0.243	0.0000008	0.993
2	0.0000152	0.155	0.377	0.0000007	0.999
3	0.0000183	0.134	0.457	0.0000009	0.999
4	0.0000186	0.0836	0.243	0.0000005	0.998
5	0.0000197	0.139	0.326	0.0000006	0.999
6	0.0000194	0.157	0.426	0.0000007	0.999
7	0.0000338	0.075	0.275	0.0000008	0.999
8	0.0000327	0.132	0.467	0.0000007	0.999
9	0.0000605	0.064	0.496	0.0000015	0.999

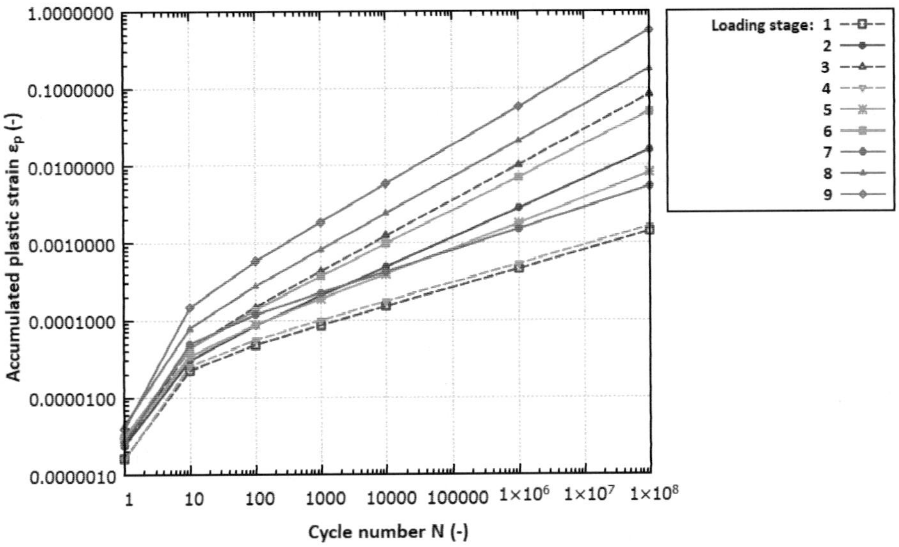

Figure 4. Plastic strain accumulation prognosis based on Eq. (3) for RAP at three different confining pressures (i) $\sigma_3 = 20.7$kPa (blue), (ii) $\sigma_3 = 35$kPa (green), (iii) $\sigma_3 = 69$kPa (red).

The empirical parameters were analysed to facilitate the mechanical parameters for practical matters. The deployed model application is limited to the presented test results. Although, it is possible to find the physical meaning of empirical constants and therefore make possible to us the model for calculation of accumulated plastic strain. Equations (4), (5) and (6) provide the basis for empirical constants calculation based on value of σ_3 and q.

$$a = \frac{1.6}{1 + \frac{q \cdot \sigma_3}{-5818}} \tag{4}$$

$$b = 0.2 \cdot e^{\frac{-16.7}{q}} \tag{5}$$

$$c = 0.15 + 0.005 \cdot (log(\sigma_3) + q) \tag{6}$$

Above presented Equations have coefficient of determination R^2 equal to 0.871, 0.751, 0.735 respectively.

4. Conclusions

In this study a cyclic loading triaxial tests were performed on Reclaimed Asphalt Pavement (RAP) to characterize the material possible application in road engineering as a subbase layer. The three levels of confining pressure were applied which allowed to study the effect of plastic strain accumulation as well as stiffness change during the loading. The test results lead to the following conclusions:

4.1. RAP as a composite material consists of mineral grains coated with asphalt. During the studies an effect of confining pressure were observed which results in higher plastic strain accumulation. The higher forces between RAP grains and particles are a cause of viscous deformations.

4.2. Stiffness degradation effect which was studied by resilient modulus M_r and degradation modulus δ analysis led to conclusion that in general the lower confining pressure the higher stiffness degradation is noted. Nevertheless, the minor σ_3 effect was also observed due to the RAP grain deformation. These findings were confirmed by visual sample inspection after the test. RAP samples during disintegration revealed that the grains were interlocked in such a way that they created agglomerates.

4.3. The plastic strain accumulation model was deployed to study the possibility of plastic strain deformation after $1\cdot10^8$ loading cycles. The results of the analysis have shown that the long-term RAP performance is dependent on confining stress where at higher σ_3 level higher accumulated plastic deformation is perceived. The practical advice is that RAP performs well at $\sigma_3 < 35$kPa and at deviator stress $q < 41.4$kPa. Such conditions are common in road subbase what is promising for future RAP application as subbase construction material. For practical applications, the developed model was supplemented by physical representation of three empirical parameters. The Eq. (4) to (6) although. using this equations requires caution.

5. References

1. Kaluder J. Cyclic load triaxial test procedures for soils in transportation engineering. IOP Conf Ser: Mater Sci Eng. 2019 Jun 1;566(1):012013.

2. Patel A, Kulkarni MP, Gumaste SD, Bartake PP, Rao KVK, Singh DN. A Methodology for Determination of Resilient Modulus of Asphaltic Concrete. Advances in Civil Engineering. 2011;2011:1–6.

3. Lim SM, Indraratna B, Heitor A, Yao K, Jin D, Albadri WM, et al. Influence of matric suction on resilient modulus and CBR of compacted Ballina clay. Construction and Building Materials. 2022 Dec;359:129482.

4. Hao S, Pabst T. Prediction of CBR and resilient modulus of crushed waste rocks using machine learning models. Acta Geotech. 2022 Apr;17(4):1383–402.

5. Sas W, Głuchowski A, Gabryś K, Szymański A. Application of cyclic CBR test for the estimation of resilient modulus in the pavement construction. Geotechnical Engineering for Infrastructure and Development. 2015 Jan;3747–52.

6. Nguyen BT, Mohajerani A. Resilient modulus of fine-grained soil and a simple testing and calculation method for determining an average resilient modulus value for pavement design. Transportation Geotechnics. 2016 Jun;7:59–70.

7. Singh AK, Sahoo JP. Rutting prediction models for flexible pavement structures: A review of historical and recent developments. Journal of Traffic and Transportation Engineering (English Edition). 2021 Jun;8(3):315–38.

8. Barksdale R D. Laboratory Evaluation of Rutting in Base Course Materials. In England: TRID; 1972. p. 161–74. Available from: https://trid.trb.org/View/138844

9. Taha R, Ali G, Basma A, Al-Turk O. Evaluation of Reclaimed Asphalt Pavement Aggregate in Road Bases and Subbases. Transportation Research Record. 1999 Jan;1652(1):264–9.

10. Mariyappan R, Palammal JS, Balu S. Sustainable use of reclaimed asphalt pavement (RAP) in pavement applications—a review. Environ Sci Pollut Res. 2023 Feb 23;30(16):45587–606.

11. Dong Q, Huang B. Laboratory Evaluation on Resilient Modulus and Rate Dependencies of RAP Used as Unbound Base Material. J Mater Civ Eng. 2014 Feb;26(2):379–83.

12. Arshad M, Ahmed MF. Potential use of reclaimed asphalt pavement and recycled concrete aggregate in base/subbase layers of flexible pavements. Construction and Building Materials. 2017 Oct;151:83–97.

13. Arshad M. Development of a Correlation between the Resilient Modulus and CBR Value for Granular Blends Containing Natural Aggregates and RAP/RCA Materials. Advances in Materials Science and Engineering. 2019 Apr 1;2019:1–16.

14. Solaimanian M, Tahmoressi M. Variability Analysis of Hot-Mix Asphalt Concrete Containing High Percentage of Reclaimed Asphalt Pavement. Transportation Research Record. 1996 Jan;1543(1):89–96.

15. Tebaldi G, Dave EV, Marsac P, Muraya P, Hugener M, Pasetto M, et al. Synthesis of standards and procedures for specimen preparation and in-field evaluation of cold-recycled asphalt mixtures. Road Materials and Pavement Design. 2014 Apr 3;15(2):272–99.

16. Alam TB, Abdelrahman M, Schram SA. Laboratory characterisation of recycled asphalt pavement as a base layer. International Journal of Pavement Engineering. 2010 Apr;11(2):123–31.

17. Standard Specification for Classification of Soils and Soil Aggregate Mixtures for Highway Construction Purposes. AASHTO M 145-91 2017.

18. Polish General Director of National Roads and Motorways. Unbound Mix for National Roads; Technical Specifications, Directive No 102 of Polish General Director of National Roads and Motorways. WT–4 2010.

19. Standard Method of Test for Determining the Resilient Modulus of Soils and Aggregate Materials. AASHTO T 307.

Analysis of the impact of rail head wear on railway superstructure stability

K Mirchev, M Lepoev

University of Architecture, Civil Engineering and Geodesy (UACEG), 1000 Sofia, Bulgaria

kmirchev_fte@uacg.bg, mlepoev_fte@uacg.bg

Abstract. Continuous welded rail track (CWR) is a superstructure construction of the railway that withstands significant compressive forces from the longitudinal extension of the rails due to temperature changes, ensuring the safe railway operation. The applicable for the Bulgarian railway system methodology for determining the critical compressive force at which the CWR loses stability has been examined. Different types of rails with their respective design parameters, important for the force calculation, have been taken into consideration. The considered design parameters, as moment of inertia and sectional modulus of the rails, change during operation due to the rail head wear. The aim of the article is to examine specific examples of railway tracks on curves, analyzing the impact of the changed operational design parameters of the rail on the stability of the track. The operational and geometric parameters of the railway track have been considered.

Keywords: Rail Head Wear, Moment of Inertia, Continuous Welded Rail (CWR), Track Stability, Critical Compressive Force

1. Introduction

The stability of a Continuous Welded Rail track (CWR) is influenced by both, the type of superstructure and the in-service condition of the railway track. During the design phase, track stability is calculated using design parameters for the track component. Once in service, these components must be maintained within certain threshold limits to ensure safe operation of the system.

In a CWR, the middle section of the rails remains fixed by the temperature changes, while dilatation occurs only at the ends of the welded rails. Consequently, significant longitudinal forces are generated in the middle section of the rail, positive in the summer and negative in the winter. The horizontal track resistance to the longitudinal compressive force is determined by the rail stiffness, the resistance of the ballast prism on the sleepers against horizontal displacement and the stiffness of the track superstructure determined by the fastening systems.

This study focuses on rail stiffness as a component of track stability and examines the impact of in-service changes in rail parameters on it. Specifically, the paper aims to analyze how rail head wear during service affects critical forces, which could lead to track instability, while considering relative track layout parameters.

2. The rail as a component of the track superstructure. Rail wear

The manufacturing requirements, qualifying tests, and acceptance tests, as well as the various types of rails with their technical parameters, are defined in [2]. The most used types of rails on the Bulgarian railway network for main tracks are the 60E1 and 49E1 (or S49) rails. The cross-section profiles of both types, along with their relevant geometrical parameters, are provided in Appendix A of [2] and shown in Figure 1.

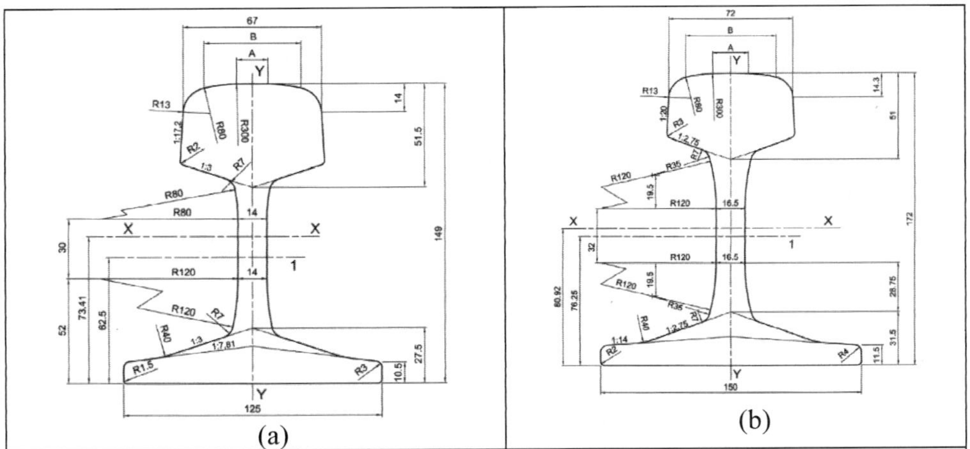

cross-sectional area: 62.92 cm^2	cross-sectional area: 76.70 cm^2
mass per meter: 49.39 kg/m	mass per meter: 60.21 kg/m
moment of inertia x-x axis: 1816 cm^4	moment of inertia x-x axis: 3038.3 cm^4
section modulus - Head: 240.3 cm^3	section modulus - Head: 333.6 cm^3
section modulus - Base: 247.5 cm^3	section modulus - Base: 375.5 cm^3
moment of inertia y-y axis: 319.1 cm^4	moment of inertia y-y axis: 512.3 cm^4
section modulus y-y axis: 51.0 cm^3	section modulus y-y axis: 68.3 cm^3

Figure 1. Rail profile parameters types 49E1 (a) and 60E1 (b) [2].

The interaction between the rails and the wheels of the rolling stock, due to tractive and guiding forces, leads to material fatigue and material abrasion, corresponding to rail head wear. This wear is typically observed on the running surface, particularly in curves at the edge of the outer rail and lips on the inner rail. In this study, we focus on investigating rail wear in curves with narrow radii, where centrifugal forces cause unsymmetrical wear between the rails, increasing the risk of track instability.

Two significant factors considered in this investigation are the cross-sectional area of the rails and the moment of inertia about the vertical axis I_{y-y}. To assess the wear, we differentiate between vertical wear of the rail head (a_v), measured in the middle of the rail head, and lateral wear (a_l), measured 14 mm below the running surface on the inner side of the rail. Although these types of rail wear are measured and evaluated separately, they are interdependent. Measurement requirements and estimation guidelines for rail wear are defined in [3].

According to [3] the maximum vertical wear of the rail head should be limited to the value at which there is a risk of interaction between the wheel flange and the joint fish bolts. In the case of CWR the maximum allowable vertical rail head wear is limited to 8 mm for rails 60E1 and 49E1. The maximum lateral rail head wear occurs when the angle between the tangent to the wear surface and the horizontal axis is 60° (Figure 2). The maximal values for lateral wear, without the presence of vertical wear, are 11 mm for rails 60E1 and 13.5 mm for rails 49E1. When both types of wear are present, the total adjusted rail head wear (a_{adj}) is calculated as:

$$a_{adj} = a_v + k.a_l \qquad (1),$$

where k=0.36 for rails heavier than 41 kg/m. The limit values for adjusted rail head wear depend on the line speed and the axis distance between the sleepers [3]. An alternative approach for measuring lateral rail head wear, as presented in Figure 2, is provided by [4]. Herein, the lateral wear shall be limited to 10 mm for speed (V km/h) V>160 km/h, 14 mm for 120<V≤160 and 16 mm for V≤120 km/h.

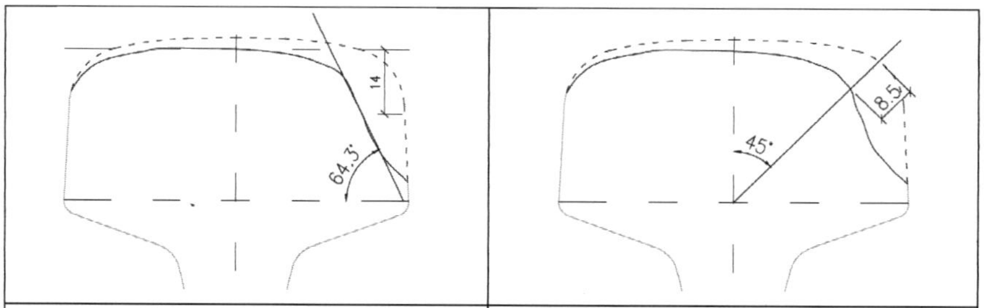

Lateral rail head wear measurement approach according to [3]	Lateral rail head wear measurement approach according to [4]

Figure 2. Rail wear lateral measurement approaches.

Continuous Welded Rail (CWR) tracks require new rails during manufacturing. However, in narrow curves with significant rail wear, it is acceptable to interchange the positions of inner and outer rails [1]. Additionally, used rails with initial rail head wear can be implemented following the requirements provided in [1] and [3].

Rail head wear is one of the factors contributing to track gauge widening, a factor for increasing the nosing of the railway vehicle on the track, which, in turn, affects the allowances for calculation of the kinematic structure gauge. Determination of the kinematic gauge, considering the in-service track parameters is studied in [5].

The implementation of CWR represents a significant change in the railway system. The possibility of introducing used rails into the track must take into consideration the operational parameters of the line, in addition to meeting the suitability requirements for the used rails. Therefore, a risk-based approach is recommended when making decisions in this regard. The application of the Common Safety Method for Risk Assessment is analyzed in [6].

3. The impact of rail head wear on rail profile parameters

According to the methodology for determining the critical compressive force for track stability provided in [1], the geometrical parameters of the rail profiles that have an impact are the cross-sectional area and the moment of inertia about the y-y axis. These parameters decrease with material abrasion.

In the categorization of different factors influencing horizontal displacement resistance, as provided by Lichtberger in Table 17 of [4], the moment of inertia of the rails is classified as "good". The categorization ranges from "very poor" to "very good". It is also stated that the moment of inertia is one of the parameters that influence the rate of track quality deterioration and is a component in the suggested model for calculating the rate of deterioration [4].

An analysis of the influence of rail head wear on the value of the moment of inertia (I) and sectional modulus (W) is presented in [7]. The study focuses on rails type 60E1 and UIC 60, with the analysis conducted on samples of used rails previously prepared for this purpose. The acquired data are processed using Autodesk Autocad (Computer Aided Design software) to calculate the real moments of inertia and sectional modulus of worn rails [7].

Another study on this topic analyzes the rail-wear limits based on fracture strength. For estimation, a methodology is used where the actual rail cross-section is approximated by an idealized section consisting of three rectangular areas representing the head, web, and base of the rail [8].

4. Determination of the CWR track stability

The rules for determining track stability in the Bulgarian railway network using continuous welded rail tracks are outlined in [1]. Document [1] represents a company rule of the Bulgarian Infrastructure Manager (NRIC) and provides a methodology for calculating the critical compressive forces (N_{cr}) and the corresponding maximum temperature at which the track loses stability. The railway network is divided into three climate zones, each having a specific temperature amplitude for rail welding, considering the track layout and the curve radius.

Formula for the calculation of N_{cr} is given in (2).

$$N_{cr} = \frac{6-K.\frac{B.l^2}{2}+\sqrt{24.K.\frac{B.l^2}{2}(1+6.K.i_r)}}{1+6.K.i_r}.\frac{E_r.l_r}{l^2} \qquad (2)$$

K is a coefficient representing the fastening resistance depending on the sleepers;

B is a coefficient representing ballast resistance on the sleeper;

l is the center distance between sleepers, calculated for the required sleepers quantity in 1 km railway track;

$i_r = \frac{E_r.I_r}{l}$ (3) is the rail stiffness.

In curves, N_{cr} is corrected with the coefficient k, considering the reduction in critical force depending on the radii.

$$N_{cr,r} = k.N_{cr} \qquad (4)$$

The estimation for the CWR track stability is given by the safety coefficient, calculated according to (5). The maximal compressive force $maxN_t$ is a function of the maximal temperature amplitude for the given climate zone ($max\Delta T$) and the longitudinal compressive force (N_o), determined for a temperature increasement of $1°C$. For new rails type 49E1 $N_o = 15.6\ kN$ and for type 60E1 $N_o = 19.05\ kN$.

$$k_s = \frac{N_{cr}}{maxN_t} \qquad maxN_t = max\Delta T.N_o \qquad N_o = E_r.A.\alpha \qquad (5)$$

The parameters module of elasticity E_r and coefficient of thermal expansion (α) are based on steel R260, A is the cross-sectional area.

Table 1 presents an example on calculated design value of critical compressive force, in straight line and in curve with radius 300m, for rails 60E1 and 49E, sleeper type CT-6 and center distance between sleepers 60 cm.

Table 1. Design value of critical compressive force.

	Sleeper type	Distance between sleepers, cm	N_{cr}, kN	k	$N_{cr,r}$, kN	$maxN_t$, kN	k_s
Straight line							
49 E1	CT-6	60	2014	1	2014	640	3.15
60 E1	CT-6	60	2578	1	2578	703	3.67
Curve R=300 m							
49 E1	CT-6	60	2014	0.53	1067	539	1.98
60 E1	CT-6	60	2578	0.53	1366	608	2.25

5. In service rail wear. Case study

The data acquisition is provided during the regular inspection of the railway network, as part of the maintenance services of NRIC. The object of the inspection is the measurement of the track geometry parameters and rail profiles measurement. The inspection is conducted with track measuring car "PLASSER & THEURER" EM-120 equipped with track geometry measurement system (TGMS) and rail profile measurement system (RPMS). The RPMS is contactless system using laser technology and scanning the full profile of the rail. The controlled parameters are vertical wear measured at the middle of the rail head and lateral wear, measured at the inner side of the rial, 14 mm below the running surface. The data for the rail head wear is recording every 25 cm. All of the systems are measured simultaneously at the same recording step, which allows correlation of the rail head wear with the track geometry and track layout parameters like curvature of the track, cant, alignment, track gauge.

IOP Conf. Series: Materials Science and Engineering 1297 (2023) 012014 doi:10.1088/1757-899X/1297/1/012014

Object of the case study are the railway intersections: Delyan – Dyakovo and Hrabursko – Razmenna. The choice is based on the following criteria: the track superstructure to be executed as CWR, the track layout to content curves with narrow radius, to have sections with different type of rails. For all investigated section the superstructure consists of prestressed monoblock concrete sleepers and elastic fastening system.

The method for determination the track stability presented with Formula (2) takes into account the parameters for one rail. Thus, only the rail with a higher rail wear rate, the outer rail, is taken into analysis. The initial input data for calculation of the critical compressive force (Formula 2) with the in-service changed rail parameters is given in Table 2.

Table 2. Information about the parameters of the tracks object of investigation.

Rail section - Delyan-Dyakovo								
Curve	Start km	End km	Radius, m	Sleepers offset, cm	Sleeper	Rails, type	Steel	Climate zone
Curve 1	73+900	74+200	~300	60	CT-6	60 E1	R 260	2nd climate zone
Curve 2	77+650	77+900	~300	60	CT-6	60 E1	R 260	2nd climate zone
Rail section – Hrabursko - Razmenna								
Curve 3	29+580	29+660	Transition curve	60	CT-6	49 E1	R 260	2nd climate zone
Curve 4	29+930	30+150	~550	60	CT-6	49 E1	R 260	2nd climate zone

By the measurement of the parameter rail head wear, data for the vertical and lateral wear of the left and the right rail, are recorded at every 25 cm. Additionally, at certain points, the full profile of the rail with the wear dimensions is depicted. That specific full profiles are used to understand the parabola of the wear. Like methodology used in [7] Autodesk AutoCAD application is used to calculate the real moment of inertia. Based on that result, the rail head is approximated by an idealized section of rectangular, applying methodology of [8]. The applied approach with the output data is depicted in Figure 3. Further, for the establishment of the in-service parameters of worn rail for given values of wear is used the idealized section. The study is focusing on the moment of inertia according to y axis I_{y-y}, where all geometrical centers are at the same plane, therefore, only the rail head is approximated to a rectangular. The approximation of the rail head cross profiles shows that the fault between results obtained by the real and the idealized profile is acceptable.

The in-service moment of inertia and cross-sectional area are analyzed at the point with maximal vertical and lateral wear and maximal adjusted wear. However, single points of extreme values significantly exceeding the surrounding values are observed. These extreme maximums at single section do not correspond to the common wear of the rail. Therefore, average wear for distance of 25 m is calculated. The length of 25 m is based on the usual length of a single rail welded into the track. As relevant for the recalculation of the critical compressive force is taken the section with a maximum average value of the adjusted rail wear.

The case study is focused on 4 sections, presenting a section of two curves from the Delyan-Dyakovo intersection and a section of two curves from the Hrabursko - Razmenna intersection. For each studied section is provided a real measured rail cross-section profile, representing the typical wear patterns, with the dimension of the vertical and the lateral wear (Figure 4 and Figure 5). Using the Autodesk AutoCAD application, the in-service cross-sectional area and the moment of inertia of the worn rails is established (Figure 4 and Figure 5). That analysis is used to approximate the dimension of the idealized profiles, where maximum average adjusted wear is obtained. Further, with the obtained data on the cross-sectional area and moment of inertia, the track stability for worn rails is recalculated.

IOP Conf. Series: Materials Science and Engineering 1297 (2023) 012014 doi:10.1088/1757-899X/1297/1/012014

The inputs data for the described approach are provided in columns 1 to 4 of Table 3, where columns 1, 2 and 3 present the average wear values above section of 25 m and column 4 provides information about the local curvature.

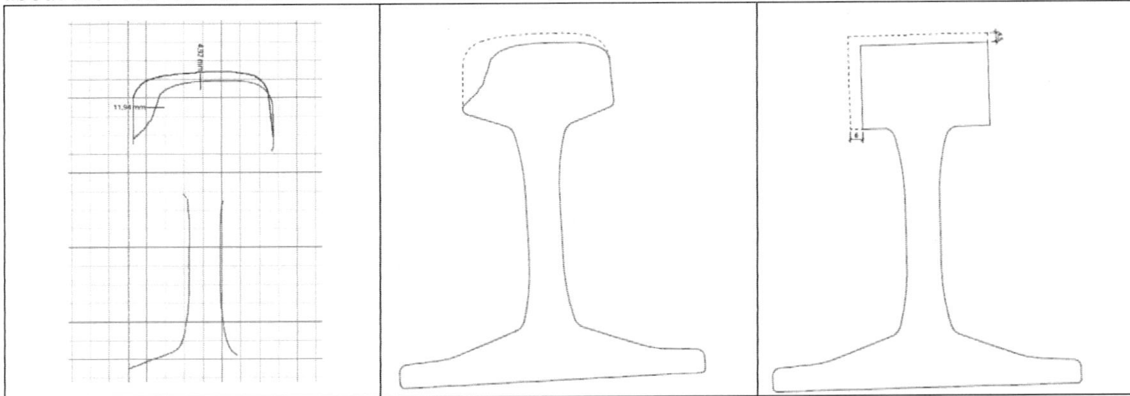

Vertical wear: 4.92 mm Lateral wear: 11.94 mm	I_{y-y} : 474.1 cm^4 A: 70.8 cm^2	I_{y-y} :475.2 cm^4 A:72.1 cm^2 Approx. vert. wear: 4.5 mm Approx. lateral wear: 6 mm

Figure 3. Rail head wear approximation.

Delyan – Dyakovo Curve 1 Rail 60 E1, km 74+102	Delyan – Dyakovo Curve 2 Rail 60 E1, km 77+843

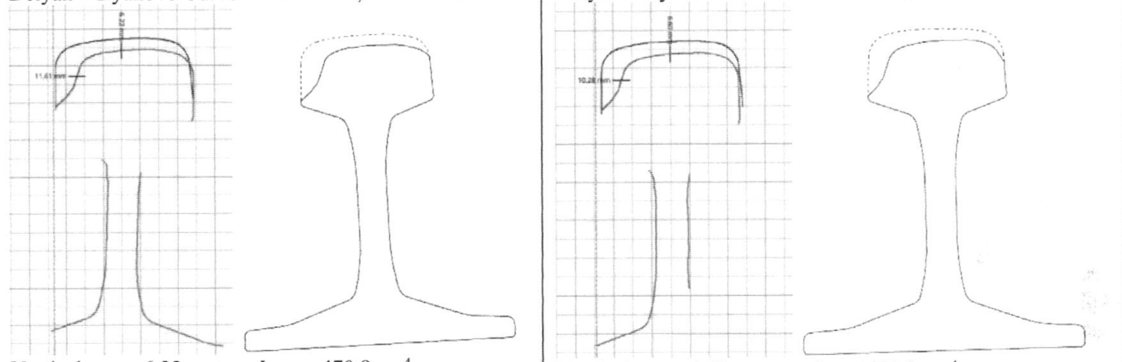

Vertical wear: 6.22 mm I_{y-y} : 470.8 cm^4 Lateral wear:11.61 mm A: 70 cm^2	Vertical wear: 5.58 mm I_{y-y}: 471.3 cm^4 Lateral wear:11.92 mm A: 69.9 cm^2

Figure 4. In-service rail head wear for rails 60 E1, railway section Delyan-Dyakovo.

Hrabursko – Razmenna Curve 3 Rail 49 E1, km 29+605	Hrabursko – Razmenna Curve 4 Rail 49 E1, km 30+040

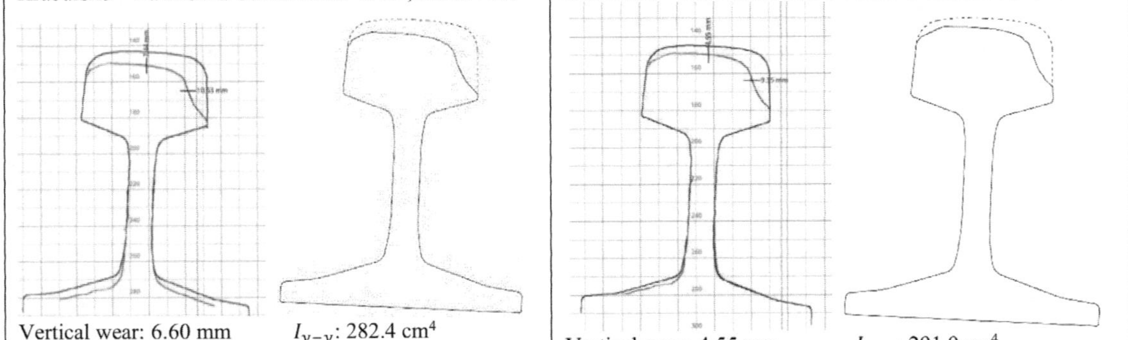

Vertical wear: 6.60 mm I_{y-y}: 282.4 cm^4 Lateral wear:10.28 mm A: 55.8 cm^2	Vertical wear: 4.55 mm I_{y-y}: 291.9 cm^4 Lateral wear:9.55 mm A: 57.9 cm^2

Figure 5. In-service rail head wear for rails 49 E1, railway section Hrabursko – Razmenna.

The established rail profile parameters (A and I_{y-y}) for worn rails given in column 5 and 6 of Table 3 are the input data for the track stability recalculation for worn rails presented in Table 4. The coefficient for correction of critical force in curve (column 4 of Table 4) is based on the established by the TGMS curvature of the track for the examined section of 25 m. The analysis for the impact of the rail head wear on the track stability is provided in Table 5, presented as a ratio expressed in % between the design value for new and the in-service value for worn rails. Negative values in column 7 mean that the safety coefficient is increased.

Table 3. Rail head wear in service.

	Adjustive, mm	Vertical, mm	Lateral, mm	Curvature, m	Alignment D 3-25 m, mm	In-service A, cm^2	In-service I_{y-y}, cm^4
	1	*2*	*3*	*4*	*5*	*6*	*7*
Curve 1	12.17	8.64	9.81	298	0,5325	69.9	467.6
Curve 2	10.36	6.63	10.37	300	3,0578	70.6	467.5
Curve 3	12.43	8.82	10.02	761	0,6924	55.2	280.3
Curve 4	12.76	9.26	9.72	541	1,5603	55.1	280.4

Table 4. Recalculation of the track stability for worn rails.

	A, cm^2	I, cm^4	**Ncr, kN**	**k**	**Ncr,r, kN**	**No, kN**	**maxNt, kN**	**ks**
	1	*2*	*3*	*4*	*5*	*6*	*7*	*8*
Design 60E1	76.7	512.3	2578	0.53	1366	19	608	2.25
Design 49E1	62.92	319.1	2014	1	2014	15.6	539	3.73
				In Service				
Curve 1 60E1	69.9	467.6	2459	0.53	1303	17.3	554	2.35
Curve 2 60E1	70.6	467.5	2458	0.53	1303	17.5	560	2.33
Curve 3 49E1	55.2	280.3	1881	1	1881	13.7	473	3.97
Curve 4 49E1	55.1	280.4	1881	1	1881	13.7	472	3.98

Table 5. Ratio expressed in % between the design value and the in-service value of worn rails.

	A, %	**I, %**	**Ncr, %**	**Ncr,r, %**	**No, %**	**maxNt, %**	**ks %**
	1	*2*	*3*	*4*	*5*	*6*	*7*
Curve 1 60E1	8.87	8.73	4.62	4.61	8.95	8.88	-4.44
Curve 2 60E1	7.95	8.74	4.65	4.61	7.89	7.89	-3.56
Curve 3 49E1	12.27	12.16	6.60	6.60	12.18	12.24	-6.43
Curve 4 49E1	12.43	12.13	6.60	6.60	12.18	12.43	-6.70

The investigated sections of lines present rail head wear profiles with clearly expressed vertical and lateral wear. Nevertheless, in practice, the rail head wear may have predominantly vertical or lateral component. In order to analyze the influence on the parameters of the rail profiles, when only one type of wear is represented, approximated rail profiles for rails 60 E1 and rails 49 E1 with a defined wear from 1 mm to 10 mm with a step of 1 mm are further investigated (Figure 6).

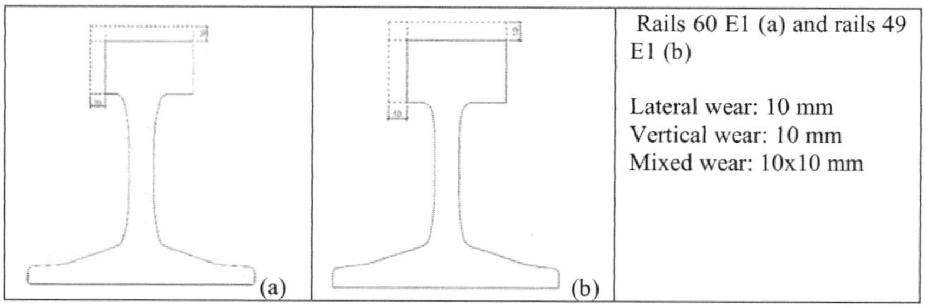

Figure 6. Approximated rail profiles with a defined wear of 10 mm.

The track stability recalculation results for approximated wear of 10 mm are presented in Table 6, while the analysis for the impact of the rail head wear on the track stability is provided in Table 7. The graphics on Figure 7 show how the rail profile parameters and the critical pressure force change with the wear rate.

Table 6. Recalculation of the track stability for different types of typical rail head wear.

	A, cm^2	**I, cm^4**	**Ncr, kN**	**k**	**Ncr,r, kN**	**No, kN**	**maxNt, kN**	**ks**
	1	*2*	*3*	*4*	*5*	*6*	*7*	*8*
Design 60E1	76.7	512.3	2578	1	2578	19	703	3.67
Design 49E1	62.92	319.1	2014	1	2014	15.6	539	3.73
Approximated wear of 10 mm for rails 60 E1								
Lateral	72,2	461,3	2441	1	2441	17.9	662	3.69
Vertical	69.5	481.7	2497	1	2497	17.2	637	3.92
Mixed	65.5	443.9	2393	1	2393	16.2	601	3.98
Approximated wear of 10 mm for rails 49 E1								
Lateral	58.2	279.6	1878	1	1878	14.4	498	3.76
Vertical	56,1	296.7	1938	1	1938	13.9	481	4.03
Mixed	52.7	265.7	1829	1	1829	13.1	452	4.05

Table 7. Ratio expressed in % between the design value and the in-service value of worn rails.

	A, %	**I, %**	**Ncr, %**	**Ncr,r, %**	**No, %**	**maxNt, %**	**ks, %**
	1	*2*	*3*	*4*	*5*	*6*	*7*
Approximated wear of 10 mm for rails 60 E1							
Lateral	5.87	9.96	5.31	5.31	5.79	5.83	-0.54
Vertical	9.39	5.97	3.14	3.14	9.47	9.39	-6.81
Mixed	14.60	13.35	7.18	7.18	14.74	14.51	-8.45
Approximated wear of 10 mm for rails 49 E1							
Lateral	7.66	12.38	6.75	6.75	7.69	7.61	-1.07
Vertical	10.84	7.02	3.77	3.77	10.90	10.76	-8.04
Mixed	16.24	16.73	9.19	9.19	16.03	16.14	-8.58

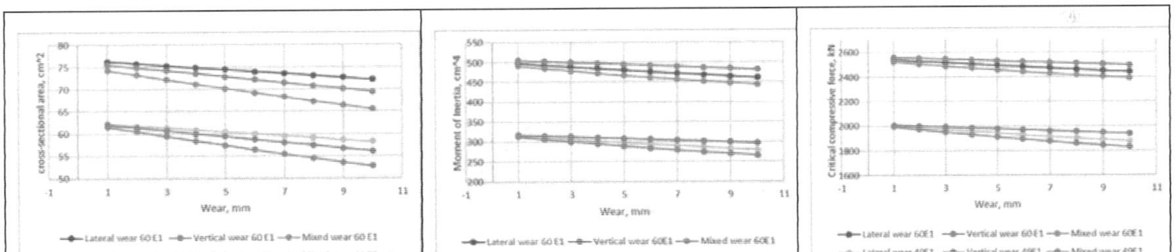

(a) (b) (c)

Figure 7. The impact of the different rail wear types to the cross-sectional area (a), the moment of inertia (b) and the critical compressive force (c).

6. Conclusion

Within the study two railway intersections, representing different types of superstructures, are examined,. Within each intersection two short section are selected, where the track is in curve with narrow radii and significant values of rail head wear are observed. Two different approaches are used to determine the parameters of worn rail profiles, one is based on the actual measured cross-sectional profiles and the other is the approximation of the wear values to an idealized rail profile. The analysis is provided for the average wear values calculated for 25 m section, where maximum adjusted wear is obtained.

The estimated adjustive wear is in the range from 10 mm up to 13 mm, as the vertical component varies from 6 mm to 9 mm, while the lateral varies from 9 mm to 10 mm. Due to the alignment of the track in curve, lateral wear is dominant. A reduction in the moment of inertia of about 8% results in a reduction of the critical compressive force by approximately 4 %. On the other hand, the reduced cross-sectional area results in a smaller longitudinal compressive force as a result of temperature changes, that affect the safety coefficient. Therefore, despite the reduced critical compressive force, applying the methodology established in [1], for the in-service rail parameter the safety coefficient is increasing.

The contribution of the different types of wear to the affected rail profile parameters is further investigated by applying defined wear from 1 mm to 10 mm with 1 mm step to an approximated profile. Mixed wear similarly affects rail cross-sectional area and moment of inertia, while the vertical wear results in a higher rate of reduction in cross-sectional area than moment of inertia, as the opposite effect is achieved by the lateral wear. Consequently, the lateral wear is more critical for the track stability. The data for 49 E1 and 60 E1 rails do not differ significantly when the initial rail profile parameters are considered.

Based on the study, the rail head wear as a single parameter is not expected to result in loss of track stability. The weaknesses of the methodology are expressed in not consideration the effect from the traffic load and the track alignment as a quality parameter representing the alignment fault of the track. These factors may significantly influence the track stability and the safety coefficient. For further analysis it is recommended to use an alternative methodology which considers all factors influencing the in-service track stability of CWR.

References

[1] 2018 Technical rules for design, construction and maintenance of continious welded rail track, NRIC, Sofia

[2] EN 13674-1 Railway applications - Track - Rail - Part 1: Vignole railway rails 46 kg/m and above

[3] 2018 Instruction for establishment and maintenance of railway superstructure and railway turnouts, NRIC, Sofia

[4] Dr. Lichtberger B 2005 *Track Compendium - Formation, Permanent Way, Maintenance, Economics* Eurailpress Tetzlaff-Hestra GmbH & Co. KG

[5] Georgiev L and Zhekov V 2021 Determination of a construction gauge allowing the movement of a railway vehicle for kinematic gauge GC, based on national construction gauge 1-CM2, AIP Conf. Proceedings 2439, https://doi.org/10.1063/5.0071089, Proceedings of the technical and scientific conference ECO Varna, ISSN (Online) 2367- 6965, ISSN (Print) 2367- 6299 – Scopus

[6] Kireva-Mihova D 2015 An Approach on Applying the CSM for Risk Evaluation and Assessment of Significant Changes of the Railway System, L International Scientific Conference on Information, Communication and Energy Systems and Technologies, Sofia, Bulgaria, ISBN: 978-619-167-182-3

[7] Makowski J , Stolarczyk Ł and Muzolf P 2019 Calculation of section modulus and moments of inertia of the 60E1 rail profile due to its head cross-section change, *Prace Naukowe Politechniki Warszawskiej* pp99-114

[8] Jeong D Y, Tang Y H and O. Orringer 1998 Estimation of Rail Wear Limits Based on Rail Strength Investigations, U.S. Department of Transportation Research and Special Programs Administration Volpe National Transportation Systems Center, Cambridge, Tufts University Mechanical Engineering Department, Medford

Study of the interaction of a modified rim profile with rail profiles used in the tram track of the city of Sofia

M Lepoev[1], **V Zhekov**[2]

[1]University of Architecture, Civil Engineering and Geodesy, 1 Hristo Smirnenski, Sofia, Bulgaria
[2]National Railway Infrastructure Company, 110 bul."Mariya Luiza", Sofia, Bulgaria

vladijekov@gmail.com

Abstract. The report examines the interaction of a modified rim profile with the rail profiles used for a tram track in the city of Sofia. The contact points, rolling radii difference and equivalent conicity were examined, graphs were drawn and the obtained values were analysed. Based on the obtained results, an analysis was made for the possibility of integrating the modified profile as the default tire profile in the city of Sofia.

1. Introduction

The report examines the interaction of a modified rim profile with the rail profiles used in the city of Sofia. The modified profile is based on the profile T81[1], which was used in the urban railway before 2018, with the main difference being a change in the radius of rounding between the conical working part and the flange.

Previous studies have shown that the T81 profile as a standard conical profile interacts better with the rail profiles used in the city of Sofia than the PRSf 2018 profile, which has designed radii in the working area [2].

Therefore, the present report examines a study of a modified profile, as we will call it T81N, with the most frequently used rail profiles in the city of Sofia - rail 49E1 [3], channel rails 60R2 [4]. Fig. 1 shows the schemes of the two tire profiles - T81 and T81N.

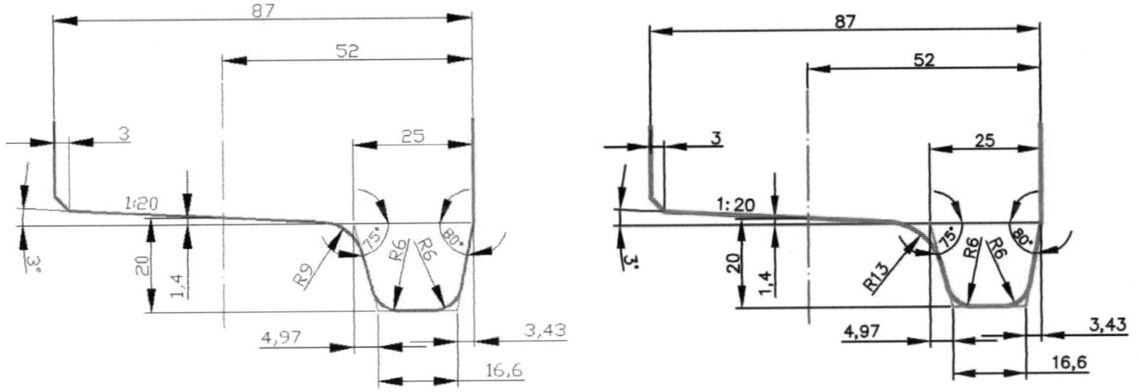

Figure 1. Rim profile T81 (left) and T81N (right)

2. Research methodology

To study the interaction of the T81N profile with the rail profiles used in the city of Sofia, the following studies were carried out:

- Contact points between rim and rail;
- Rolling radii difference;
- Equivalent conicity.

The methodology of standard EN 15302:2022 [5] was used to determine the specified characteristics, and the results were validated.

In addition, geometric studies of the contact were made to establish the interaction between the profile and the rail in certain contact positions.

3. Investigation of the contact between T81N and rail profies 49E1 and 60R2 at 1009 gauge

3.1. Contact interaction research

To establish the interaction between the T81N profile and the main rail profiles, graphs of the contact points were made. Figure 2 shows graphs of the contact points.

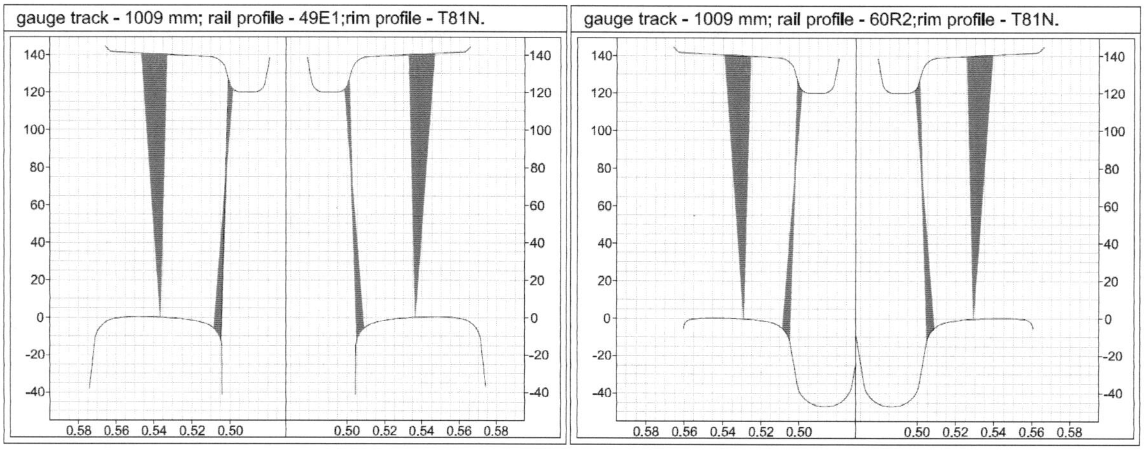

Figure 2. Contact point between T81N and rail profiles 49E1 and 60R2 at 1009 gauge

From the graph, we find that in the studied interval of offset from the axis ±10 mm there is contact between the middle part of the rail head with the conical part of the rim, as well as contact between the flange of the rim and the end of the rail. If we examine the contact position between the rim and the rail near the contact with the flange, we get the following:

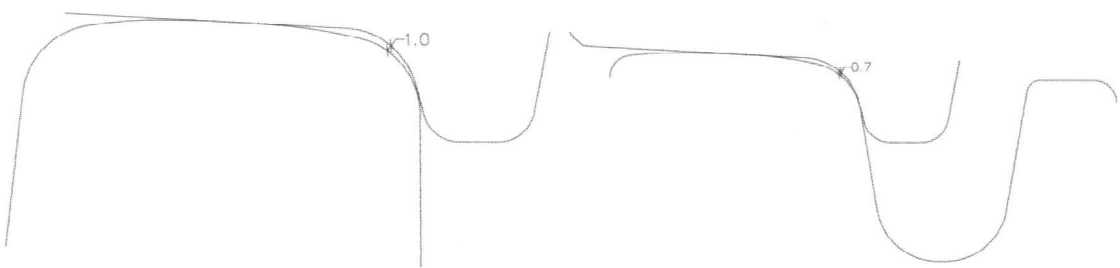

Figure 3. Geometrical analysis between T81N and rail profiles 49E1 and 60R2 at 1009 gauge

As it can be seen from the graph, the gap in both profile 49E1 and 60R2 is less than 1mm, and there is a clear two-point contact. We can say that this is favorable as it is close to a conformal contact, with the 13 m radius rounding contributing to this result. With a standard T81 profile, the gap values are higher.

3.2 Rolling radii difference(RRD)

To study the difference in the rolling radii for profile T81 N with rail profiles 49E1 and 60R2, a graphical study was made, as shown in Fig.4.

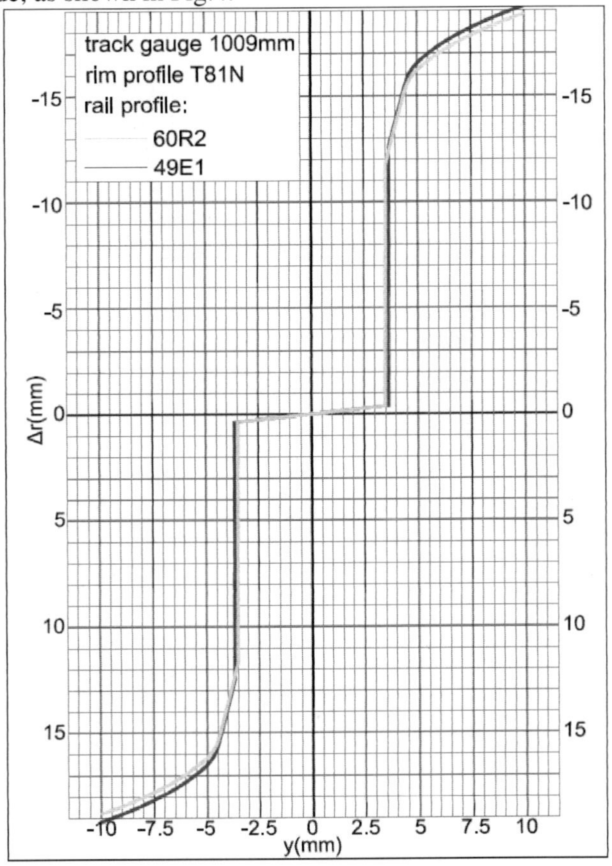

Figure 4. RRD at T81N and rail profiles 49E1 and 60R2 at 1009 gauge

We find a rise of the rail at ~3.5mm, with a smooth rise with a flat 1:20 slope, characteristic of the standard conical profile. The values are optimal and show good interaction.

3.3. Equivalent conicity

Calculations have been made for equivalent conicity in case of T81N rim profile in interaction with rail profiles 49E1 and 60R2. The results are shown in Fig. 5, as well as in tabular form in Table 1.

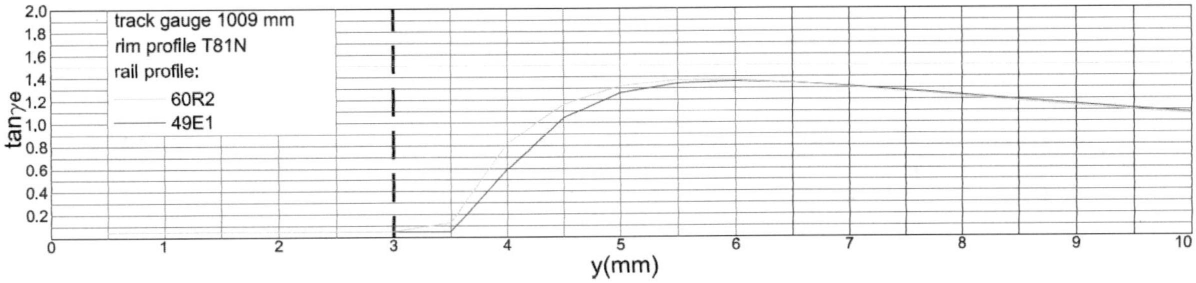

Figure 5. Equivalent conicity at rim profiles T81N with 60R2 and 49E1 – gauge 1009 mm

Table 1. Equivalent conicity at rim profiles T81N with 60R2 and 49E1 – gauge 1009 mm

Rail profile	T81N	
	y mm	$\tan\gamma_e$
60 R2	3	0.052
49E1	3	0.052

From the attached data, we find that the equivalent conicity for both profiles is in optimal values corresponding to a slope of 1: 19.2.

4. Investigation of the contact between T81N and rail profiles 49E1 and 60R2 at 1435 gauge

4.1. Contact interaction research

To establish the interaction between the T81N profile and the main rail profiles, graphs of the contact points at a gauge of 1435 mm were made. Figure 6 shows graphs of the contact points.

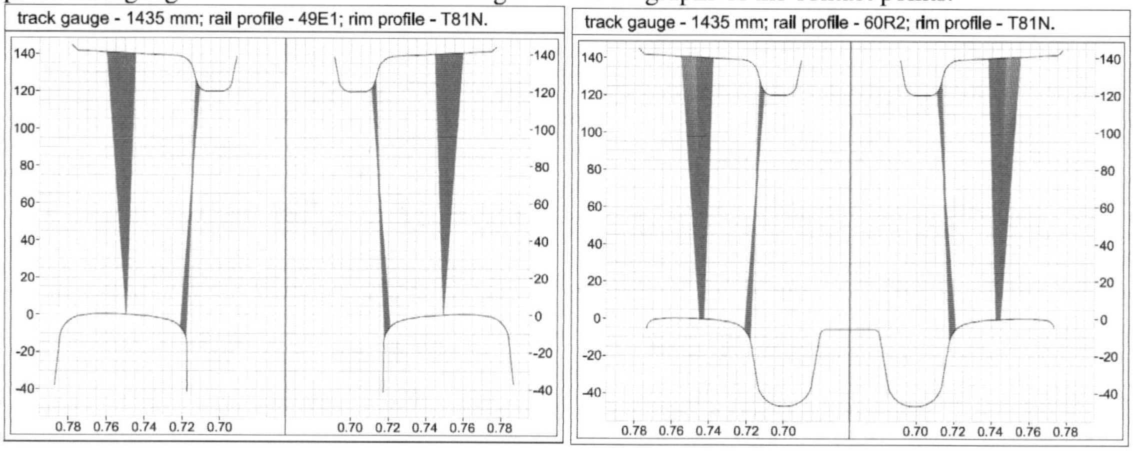

Figure 6. Contact point between T81N and rail profiles 49E1 and 60R2 at 1435 gauge

From the graph it can be seen that the contact is again mainly at the middle part of the rail and the conical part of the rim, as well as with the flange of the rim. In order to establish the gap in the case of adjacent profiles, geometrical analyzes were made.

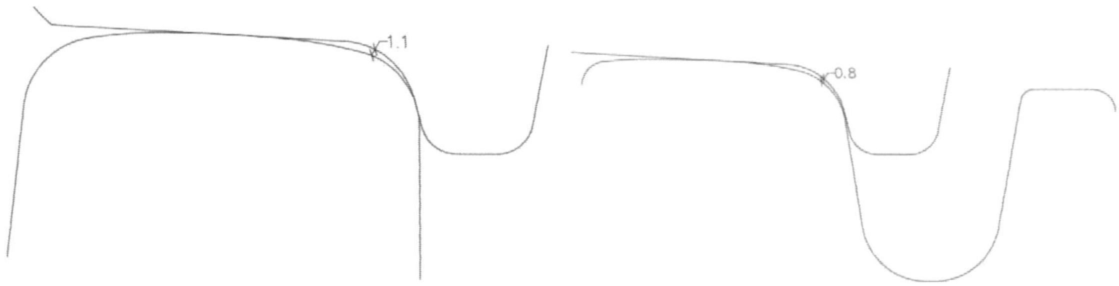

Figure 7. Geometrical analysis between T81N and rail profiles 49E1 and 60R2 at 1435 gauge

As can be seen from the graph, the gap in both profiles is close to 1mm, with a clear two-point contact. We can say that this is favorable as it is close to a conformal contact, with the 13 m radius rounding contributing to this result. With a standard T81 profile, the gap values are higher.

4.2. Rolling radii difference

To study the difference in the rolling radii for profile T81 N with rail profiles 49E1 and 60R2, a graphical study was made, as shown in Fig.8.

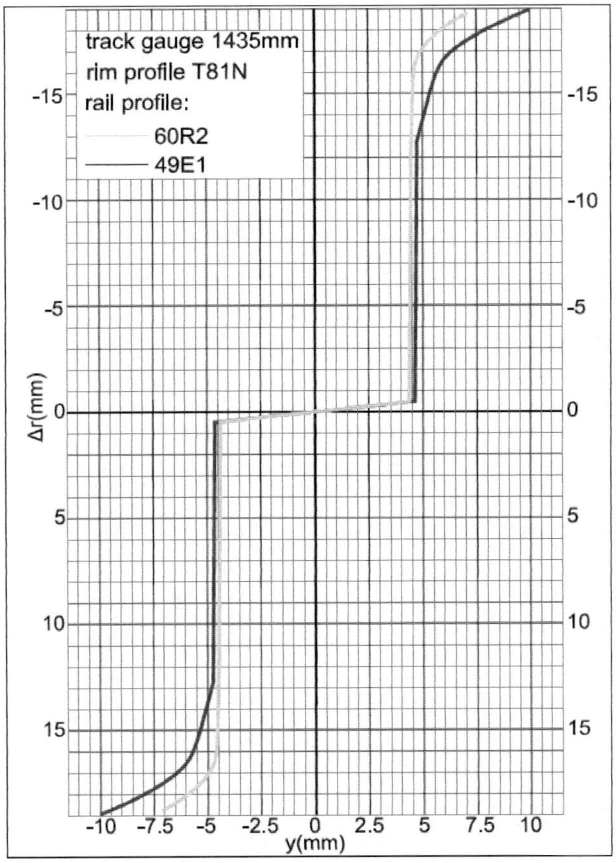

Figure 8. RRD at T81N and rail profiles 49E1 and 60R2 at 1435 gauge

We find rail rise at ~4.5mm, with a smooth rise with a flat 1:20 pitch, typical of the standard conical profile. The values are optimal and show good interaction.

4.3. *Equivalent conicity*

Calculations have been made for equivalent conicity in case of T81N rim profile in interaction with rail profiles 49E1 and 60R2. The results are shown in Fig. 9, as well as in tabular form in Table 2.

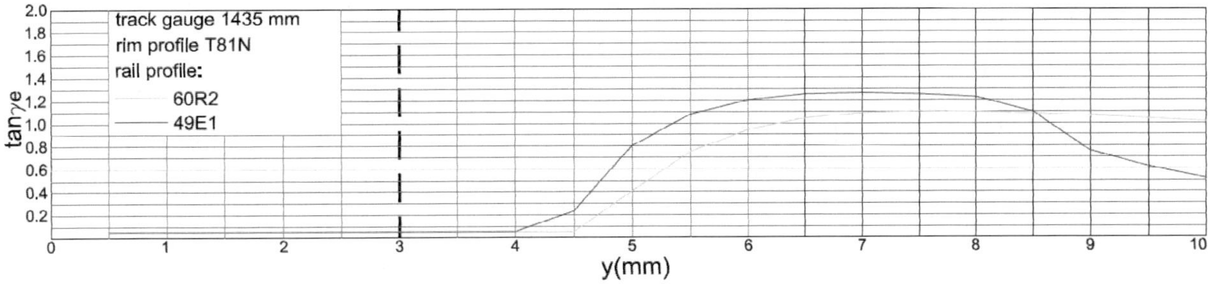

Figure 9. Equivalent conicity at rim profiles T81N with 60R2 and 49E1 – gauge 1435 mm

Table 2. Equivalent conicity at rim profiles T81N with 60R2 and 49E1 – gauge 1435 mm

Rail profile	T81N	
	y mm	$\tan\gamma_e$
60 R2	3	0.052
49E1	3	0.054

From the attached data, we find that the equivalent conicity for both profiles is in optimal values corresponding to the slope of the conicity 1: 18.5÷19.2.

5. Conclusion

From the research done, it can be clearly concluded that T81N rim profile interacts well with the most used rail profiles in the city of Sofia - 49E1 and 60R2, and the values of RRD, equivalent conicity are optimal. In the geometric tests, it is found that the creep of radius 13 reduces the gap between the rail and the rim, with which the contact is two-point, but with values of about 1 mm of the gap close to conformal. From the above, it can be concluded that profile T81N is an optimal rim profile and can be further investigated for future test application.

References

[1] Zhekov, V., Analysis of the possibilities to improve the durability of the construction of the urban railway, PHD Thesis. Defended on 20.04.2018г., VTU "Todor Kableshkov ", Sofia

[2] Zhekov, V., Study of the contact between design profiles of rails and rims used in the tram track of the city of Sofia, XI International Conference "Heavy Machinery-HM 2023", Vrnjačka Banja, 21–24 June 2023, ISBN-978-86-82434-01-6

[3] CEN European Committee for Standardization, EN 13674-1:2011. Railway applications – Track – Rail, Part 1: Vignole railway rails 46 kg/m and above, 2011

[4] CEN European Committee for Standardization, EN 14811:2006+A1:2010. Railway applications - Track - Special purpose rail - Grooved and associated constructio , 2011

[5] CEN European Committee for Standardization, EN 15302:2022. Railway applications. Wheel-rail contact geometry parameters. Definitions and methods for evaluation

[6] Allen P., Bevan A., Determination of Tramway Wheel and Rail Profiles to Minimise Derailment. ORR, 2008

[7] Emil Mihaylov, Dobrinka Atmadzhova, „Study on wheel profile of tram in operation", The first international symposium for student swith papers from mechanical engineering SRMA 2011, Kraljevo, Serbia, 2011 HEAVY MACHINERY - HM 2011, c. 49-54, ISBN 978-86- 82631-58-3

Strategic development of the national road network

Desislava Paunova , Dina Tzonevska

d.paunova@api.bg , d.conevska@api.bg

Abstract. Strategically developed transport infrastructure is a basic prerequisite for efficient, effective, and sustainable transport. The transport infrastructure is essential for economic growth, for the mobility of labor resources, for the supply of transport services and for the competitiveness of the bulgarian economy. A major problem identified in relation to the characteristics and qualities of the infrastructure along the main routes is the lack of continuous, consistent, and permanent transport networks to ensure fast and safe movement over longer distances. The development of the national road network must follow the principles of continuity and synergy with the national and European strategic documents. The main strategic priorities in the development of transport are effective maintenance and modernization of the transport infrastructure, as well as increasing the security and safety of the transport system.

1. Introduction

A well-developed road network is crucial for the economic, social, and overall development of a country. It serves as a lifeline for transportation, connecting people, goods, and services across different regions. In the case of Bulgaria, a country located in Southeast Europe, a well-functioning road network is of paramount importance due to its strategic geographical location and its potential to enhance trade, tourism, and connectivity within the country and with neighboring countries.

Bulgaria's road network plays a vital role in facilitating domestic and international trade, supporting tourism, and improving accessibility to various regions. It serves as a backbone for the transportation of goods, connecting industrial centers, ports, and markets. Additionally, a well-developed road network enhances connectivity between urban and rural areas, improving access to essential services such as healthcare, education, and employment opportunities.

Currently, Bulgaria's national road network consists of a comprehensive network of highways, expressways, and regional roads. The national road network is managed and maintained by the Road Infrastructure Agency (RIA), which is responsible for planning, construction, and maintenance activities.

While significant progress has been made in recent years, there are still challenges that need to be addressed. The existing road infrastructure faces issues such as congestion, inadequate capacity, and insufficient maintenance. Some regions, particularly rural areas, experience limited connectivity and poor road conditions, hindering economic development and social integration.

Efforts have been made to improve the national road network in Bulgaria. Notable projects include the construction and expansion of major highways, such as the Struma Highway, Hemus Highway and many more, which have enhanced connectivity between major cities and regions. Additionally, investments have been made to upgrade and modernize existing roads, improving safety and efficiency.

However, despite these advancements, there is a need for further strategic development of the national road network in Bulgaria. This includes addressing existing infrastructure gaps, improving road

IOP Conf. Series: Materials Science and Engineering 1297 (2023) 012016 doi:10.1088/1757-899X/1297/1/012016

quality and safety standards, and ensuring sustainable and environmentally friendly practices in road construction and maintenance.

In this article, we will explore the strategic development of the national road network in Bulgaria, examining the importance of a well-developed road network and providing insights into the current state of the infrastructure. We will also delve into the challenges and opportunities for further development, discussing policy frameworks, funding mechanisms, stakeholder engagement, and future recommendations for a sustainable and efficient road network in Bulgaria (Tsigdinos, 2020).

2. Historical overview of road development in Bulgaria

The history of road development in Bulgaria dates back centuries. In ancient times, the region that is now Bulgaria was part of the Roman Empire, and the Romans constructed a network of well-paved roads, known as Via Militaris, to connect important cities and military outposts. These roads played a crucial role in facilitating trade and communication within the empire.

Significant developments in Bulgaria's road network occurred during the 19th and 20th centuries. In the late 19th century, after gaining independence from the Ottoman Empire, Bulgaria began investing in road infrastructure to support economic growth and national integration. The construction of the first modern roads, such as the Sofia-Varna and Sofia-Belgrade roads, commenced during this period.

In the early 20th century, Bulgaria witnessed further expansion of its road network. The government initiated the construction of new roads and the improvement of existing ones to enhance connectivity between major cities and regions. The focus was on connecting Sofia, the capital, with other important economic and administrative centers.

During the communist era (1944-1989), road development in Bulgaria received significant attention. The government invested heavily in infrastructure projects, including the construction of highways and expressways. Notable projects during this period included the Trakiya Highway, connecting Sofia with Burgas, and the Hemus Highway, linking Sofia with Varna.

Despite the progress made in road development, Bulgaria has faced several challenges and limitations. One of the main challenges has been the limited financial resources available for infrastructure investment. Insufficient funding has hindered the timely completion of road projects and the maintenance of existing infrastructure.

Geographical factors have also posed challenges. Bulgaria's mountainous terrain and diverse landscapes make road construction and maintenance more complex and costly. Harsh weather conditions, such as heavy snowfall and flooding, further impact road conditions and require additional maintenance efforts.

Additionally, inadequate planning and coordination have led to inefficiencies in road development. Lack of proper strategic planning, environmental impact assessments, and stakeholder engagement have sometimes resulted in suboptimal road designs and negative environmental consequences.

Furthermore, bureaucratic hurdles have affected the efficiency and transparency of road development processes. These issues have led to delays, cost overruns, and a lack of public trust in the road infrastructure sector. In recent years, efforts have been made to address these challenges and improve road development practices in Bulgaria. Strategic planning, increased transparency, and the adoption of sustainable and environmentally friendly approaches are being emphasized to ensure the efficient and effective development of the national road network.

3. Importance of strategic development

3.1. Economic benefits of a well-developed road network

A well-developed road network brings numerous economic benefits to our country. Firstly, it enhances connectivity and accessibility, allowing for efficient movement of goods and services. This facilitates trade and commerce, enabling businesses to reach markets more easily and efficiently. Improved connectivity also attracts investments, as companies are more likely to establish operations in areas with good transportation infrastructure (Pricing European Transport Systems).

A well-connected road network reduces transportation costs, benefiting both businesses and consumers. It enables the efficient movement of goods, reducing logistics expenses and making products more affordable. Additionally, it improves supply chain efficiency, enabling just-in-time deliveries and reducing inventory costs. Furthermore, a well-developed road network stimulates economic growth and regional development. It connects rural areas to urban centers, promoting the flow of resources, labor, and services. This helps to reduce regional disparities and create employment opportunities in previously isolated regions.

3.2. Social and environmental impacts of strategic road development

Strategic road development has significant social impacts. It improves accessibility to essential services such as healthcare, education, and emergency services. People in remote areas gain better access to hospitals, schools, and other vital facilities, enhancing their quality of life. Additionally, improved road connectivity promotes social integration by facilitating travel and interaction between different communities.

However, road development also has environmental impacts that need to be carefully managed. Construction activities can lead to habitat fragmentation, soil erosion, and loss of biodiversity. Increased road traffic can contribute to air and noise pollution, affecting the health and well-being of nearby communities. Therefore, strategic road development should consider environmental sustainability, incorporating measures to mitigate negative impacts and promote eco-friendly practices (european transport safety council, 2021).

3.3. Role of strategic planning in optimizing road infrastructure investments

Strategic planning plays a crucial role in optimizing road infrastructure investments. It ensures that resources are allocated efficiently and effectively, maximizing the benefits derived from road development projects. Strategic planning involves assessing current and future transportation needs, identifying priority areas for investment, and setting clear objectives and targets (Ministry of transport and communications, 2017).

By conducting thorough analysis and feasibility studies, strategic planning helps identify the most cost-effective and impactful road projects. It considers factors such as traffic volume, population density, economic potential, and environmental considerations. This enables decision-makers to prioritize projects that offer the highest return on investment and align with broader national development goals.

Strategic planning also promotes coordination and collaboration among various stakeholders, including government agencies, private sector entities, and local communities. It ensures that different perspectives and interests are considered, leading to more informed and inclusive decision-making processes (Technical Committee Evaluation, 2010).

Moreover, strategic planning allows for long-term thinking and anticipates future needs and challenges. It enables the identification of emerging trends, such as technological advancements and changing transportation patterns, and helps incorporate them into road development plans. This ensures that investments are future-proofed and can adapt to evolving demands.

Strategic development of the national road network is essential for realizing economic benefits, addressing social needs, and minimizing environmental impacts. By considering the economic, social, and environmental aspects of road development, strategic planning optimizes infrastructure investments, promotes sustainable growth, and enhances the overall well-being of the country and its citizens.

4. Current challenges and opportunities

4.1. Assessment of the current state of the national road network

To effectively strategize the development of the national road network, it is crucial to assess the current state of the infrastructure. This assessment provides insights into the strengths, weaknesses, and areas that require improvement. Some key aspects to consider in the assessment include:

- Road condition: Evaluate the overall condition of the roads, including pavement quality, signage, and safety features. Identify areas that require maintenance or rehabilitation.
- Capacity and congestion: Analyze traffic flow and congestion levels on major highways and urban areas. Identify bottlenecks and areas where capacity expansion is necessary.
- Connectivity: Assess the connectivity between different regions and the accessibility of remote areas. Identify gaps in the road network that hinder efficient transportation and economic development.
- Safety: Evaluate road safety measures and accident statistics. Identify areas with high accident rates and implement measures to improve safety, such as better signage, road markings, and traffic calming measures (Pricing European Transport Systems).

4.2. Strategic infrastructure objects in Bulgaria

Figure 1 Strategic infrastructure objects in Bulgaria

Identifying the strategic infrastructure objects in the national road network is crucial for effective strategic development, those mentioned above are part of them.

4.3. Opportunities for improvement and expansion

Despite the challenges, there are several opportunities for improvement and expansion of the national road network in Bulgaria:

- Strategic planning: Develop a comprehensive strategic plan that identifies priority areas for investment, considering economic potential, traffic volume, and regional development needs.
- Public-private partnerships (PPPs): Explore opportunities for PPPs to attract private sector investment and expertise in road development projects. PPPs can help bridge the funding gap and ensure efficient project delivery.

IOP Conf. Series: Materials Science and Engineering 1297 (2023) 012016 doi:10.1088/1757-899X/1297/1/012016

- Sustainable practices: Incorporate sustainable practices in road construction and maintenance, such as using eco-friendly materials, implementing energy-efficient lighting, and promoting green infrastructure.
- Technology integration: Embrace technological advancements, such as intelligent transportation systems, to improve traffic management, enhance road safety, and optimize road capacity.
- Regional cooperation: Collaborate with neighboring countries to improve cross-border connectivity and facilitate regional trade and tourism.
- Stakeholder engagement: Involve local communities, businesses, and civil society organizations in the decision-making process to ensure their needs and concerns are considered.

By addressing the identified challenges and capitalizing on the available opportunities, Bulgaria can strategically develop its national road network, improving connectivity, promoting economic growth, and enhancing the overall quality of transportation infrastructure in the country.

5. Strategic priorities for road network development

5.1. Identification of strategic goals and objectives

To guide the development of the national road network in Bulgaria, it is essential to identify clear strategic goals and objectives. These goals should align with broader national development priorities and address the specific needs and challenges of the country. Some strategic goals and objectives could include:

Improve the connectivity between major cities, regions, and neighboring countries to facilitate trade, tourism, and social integration. Another goal is to develop road infrastructure that supports economic activities, attracts investments, and stimulates regional development. To ensuring safety should be prioritize road safety measures to reduce accidents and fatalities, improving the overall safety of road users. Incorporate sustainable practices in road development, minimizing environmental impacts and promoting energy efficiency. Ensure that all citizens have access to essential services, such as healthcare, education, and emergency services, by improving road connectivity to remote areas.

5.2. Prioritization of road projects based on economic, social, and environmental criteria

To effectively allocate resources and prioritize road projects, it is crucial to consider economic, social, and environmental criteria. This ensures that investments are made in projects that offer the highest overall benefit to the country. Some criteria to consider for prioritization include: Economic impact: Assess the potential economic benefits of road projects, such as increased trade, job creation, and improved productivity. Prioritize projects that have a significant positive impact on the economy; Social impact: Consider the social benefits of road projects, such as improved accessibility to essential services, reduced travel times, and enhanced social integration. Prioritize projects that address social needs and reduce regional disparities; Environmental impact: Evaluate the environmental consequences of road projects, including habitat fragmentation, air and noise pollution, and carbon emissions. Prioritize projects that minimize negative environmental impacts and promote sustainability; Cost-effectiveness: Assess the cost-effectiveness of road projects, considering factors such as construction and maintenance costs, expected traffic volume, and long-term benefits. Prioritize projects that offer the highest return on investment (Calculating Transport Congestion and Scarcity Costs, 2008).

By using a comprehensive evaluation framework that considers these criteria, decision-makers can prioritize road projects that align with strategic goals and maximize the overall benefit to society.

6. Outlook and recommendations

6.1. Anticipated future trends and challenges in road network development

In anticipating the future trends and challenges in road network development in Bulgaria, several factors should be considered:

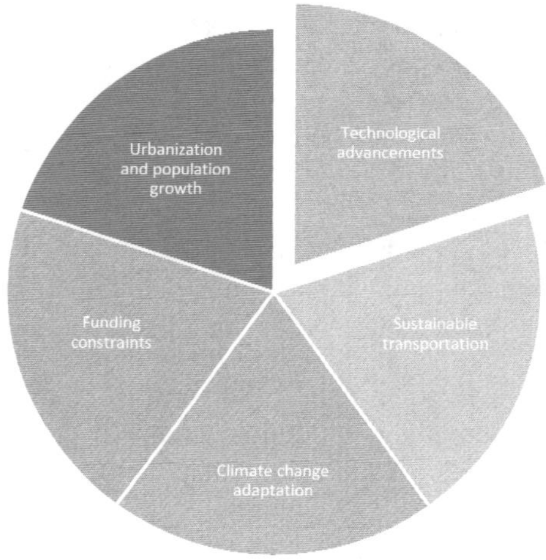

Figure 2 Challenges in road network development

Technological advancements: The rapid advancement of technology, such as electric and autonomous vehicles, will impact road network development. Infrastructure needs to be adaptable to accommodate these emerging technologies.

Sustainable transportation: There is a growing emphasis on sustainable transportation, including the promotion of public transportation, cycling, and walking. Future road network development should prioritize these modes of transportation to reduce congestion and environmental impacts.

Climate change adaptation: Climate change will pose challenges to road infrastructure due to increased frequency and intensity of extreme weather events. Future road development should incorporate climate resilience measures to ensure infrastructure can withstand these challenges.

Funding constraints: Limited financial resources will continue to be a challenge for road network development. Innovative funding mechanisms, such as public-private partnerships and alternative financing models, should be explored to bridge the funding gap.

Urbanization and population growth: The increasing urbanization and population growth in Bulgaria will put pressure on the road network. Future development should focus on improving urban mobility and addressing congestion in major cities (European commission, 2022).

6.2. Recommendations for strategic planning and development of the national road network in Bulgaria

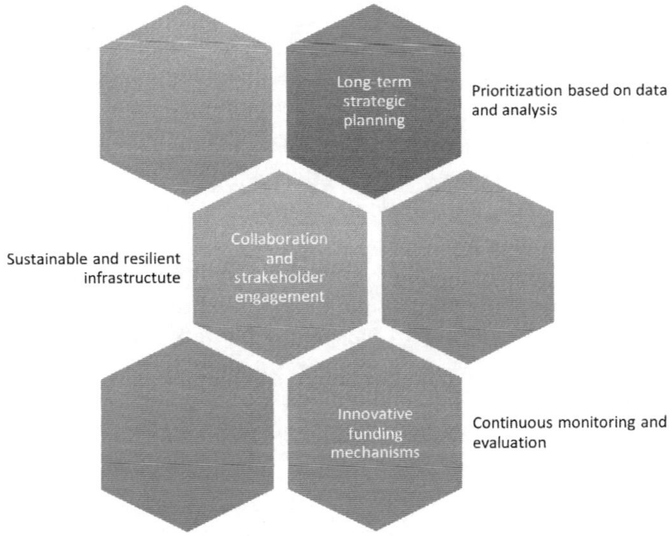

Figure 3 Strategic planning and development of the national road network

To effectively plan and develop the national road network in Bulgaria, the following recommendations should be considered:

Long-term strategic planning: Develop a comprehensive long-term strategic plan that aligns with national development goals and considers future trends and challenges. This plan should guide decision-making and resource allocation for road network development.

Prioritization based on data and analysis: Use data-driven analysis to prioritize road projects based on economic, social, and environmental criteria. This ensures that investments are made in projects that offer the highest overall benefit to the country.

Collaboration and stakeholder engagement: Foster collaboration among government agencies, private sector entities, and local communities. Engage stakeholders in the decision-making process to ensure their needs and concerns are considered.

Sustainable and resilient infrastructure: Integrate sustainable and climate-resilient practices in road network development. This includes incorporating eco-friendly materials, promoting energy efficiency, and designing infrastructure that can withstand climate change impacts.

Innovative funding mechanisms: Explore alternative funding mechanisms, such as public-private partnerships and tolling systems, to secure additional financial resources for road development projects.

Continuous monitoring and evaluation: Implement a robust monitoring and evaluation system to assess the performance and impact of road projects. This allows for adjustments and improvements to be made based on lessons learned.

By implementing these recommendations, Bulgaria can strategically plan and develop its national road network, ensuring sustainable and efficient transportation infrastructure that supports economic growth, enhances connectivity, and improves the overall quality of life for its citizens.

7. Conclusion

In conclusion, the development of the national road network in Bulgaria is crucial for promoting economic growth, improving connectivity, and enhancing the overall quality of transportation infrastructure. However, there are several challenges that need to be addressed, such as insufficient funding, inadequate maintenance, congestion, and regional disparities.

To overcome these challenges and capitalize on opportunities for improvement and expansion, it is important to conduct a thorough assessment of the current state of the road network, identify key challenges and bottlenecks, and explore opportunities for improvement. Strategic priorities should be identified, including clear goals and objectives, prioritization of road projects based on economic, social, and environmental criteria, and integration of sustainable practices in road network planning.

Anticipating future trends and challenges, such as technological advancements, sustainable transportation, climate change adaptation, and funding constraints, is essential for effective strategic planning. Recommendations for the strategic planning and development of the national road network in Bulgaria include long-term planning, data-driven prioritization, collaboration and stakeholder engagement, sustainable and resilient infrastructure, innovative funding mechanisms, continuous monitoring and evaluation, and regional cooperation (European commission, 2020).

By implementing these recommendations, Bulgaria can strategically develop its road network, ensuring sustainable and efficient transportation infrastructure that supports economic growth, enhances connectivity, and improves the overall quality of life for its citizens. It is crucial to prioritize the long-term benefits and consider the needs of all stakeholders to create a road network that meets the demands of the future and contributes to the sustainable development of the country (World Bank, 2017).

References

[1]European commission. (2022). https://europa.eu/next-generation-eu/index_bg.
Retrieved from https://europa.eu.
Pricing European Transport Systems. (n.d.).

[2]Tsigdinos, S. (2020). Strategic road network formulation: developing an alternative methodology towards sustainable mobility. Transportation Research Procedia.

[3]European comission. (2008). Guide to the analysis of investment projects by costs and benefits. Brussels: European Commission.

[4]European commission. (2020). europe.eu. Retrieved from https://trimis.ec.europa.eu/project/cooperative-vehicle-infrastructure-systems.

[5] Ministry of Transport and Communications. (2017). Integrated transport strategy in the period until 2030. Sofia: MTC.

[6]Technical Committee Evaluation. (2010). *Financing of Road Infrastructures: Guide for New Methods of Financing and Public/Private Partnership.*

[7] Calculating Transport Congestion and Scarcity Costs. (2008). *Final Report of the Expert Advisors to the High Level Group on Infrastructure Charging.*

[8]European commission. (2022). *https://europa.eu/next-generation-eu/index_bg.* Retrieved from https://europa.eu.european transport safety council. (2021, August 1st). *etsc.eu/eu-road-deaths-down-by-3900-in-2020/.* Retrieved from etsc.eu: https://etsc.eu/eu-road-deaths-down-by-3900-in-2020/

[9]OECD. (2021). *oecd.org.* Retrieved from https://www.oecd-ilibrary.org/?_ga=2.188167914.397094570.1627217129-1162414499.1627217129. *Pricing European Transport Systems.*

National transport infrastructure conference with international participation IOP Publishing
IOP Conf. Series: Materials Science and Engineering 1297 (2023) 012017 doi:10.1088/1757-899X/1297/1/012017

THE ECONOMIC FACTOR AS A KEY IN PLANNING ACTIVITIES OF THE NATIONAL ROAD NETWORK

Phd Dina Tzonevska

d.conevska@api.bg

Abstract. Transport has a key role in economic and social development. It generates economic growth by increasing access to health, education, and other national and regional public services. On one hand, investment in the transport sector boosts economic growth by increasing the return on private investment. On the other hand, the improvement of the transport service helps to reduce the prices of resources and thus - to reduce the production costs, to increase the possibilities of access to different markets. An important problem in defining the need for the construction of transport infrastructure objects is consideration of the economic factor of the projects. It is necessary for making decisions on the implementation of transport policy at the national and international level. Bringing out the economic factor as a key factor in the planning of activities on the national road network has a wide public significance, this necessitates the improvement of coordination in terms of its implementation.

1. Introduction

The road network in Bulgaria serves as a crucial infrastructure component, facilitating the movement of people, goods, and services throughout the country. It plays a pivotal role in supporting economic activities, regional development, and enhancing the overall quality of life for Bulgarian citizens. However, the planning and development of an efficient road network requires careful consideration of various economic factors.

The purpose of this article is to delve into the economic factors that significantly influence the planning activities of the road network in Bulgaria. By understanding these factors, policymakers, urban planners, and stakeholders can make informed decisions to ensure the road network's effectiveness, sustainability, and contribution to the country's economic growth.

The article will explore the multifaceted nature of economic factors and their impact on road network planning. It will analyze the cost of road construction and maintenance, the expected traffic volume, the relationship between economic development and road network planning, and the availability of funding. Each of these factors plays a crucial role in shaping the road network's design, capacity, and overall success.

Furthermore, the article will highlight the challenges and opportunities faced by Bulgaria's economy in relation to road network planning. It will address issues such as poverty, corruption, and limited funding, which can hinder the efficient development of the road network. Additionally, it will shed light on the potential economic benefits that can be achieved through strategic road network planning, including increased trade opportunities, regional development, and attracting investments.

By examining these economic factors and their implications, this article aims to provide valuable insights into the planning activities of the road network in Bulgaria. It will serve as a resource for policymakers, urban planners, and researchers, enabling them to make informed decisions and develop strategies that align with the country's economic goals.

The road network in Bulgaria is a critical component of the country's infrastructure, connecting people, goods, and services. Economic factors play a pivotal role in the planning activities of the road network, influencing decisions related to cost, traffic volume, economic development, and funding availability. By understanding and addressing these economic factors, Bulgaria can develop a robust and sustainable road network that supports economic growth, regional development, and enhances the overall well-being of its citizens.

2. Cost as an economic factor

The cost of road construction and maintenance is a significant economic factor that greatly influences the planning activities of the road network in Bulgaria. It is essential to carefully consider the financial implications of road projects to ensure their feasibility, sustainability, and long-term economic benefits (Calculating Transport Congestion and Scarcity Costs, 2008).

2.1. Analysis of the Cost of Road Construction and Maintenance in Bulgaria

Road construction and maintenance costs in Bulgaria encompass various elements, including land acquisition, materials, labor, and equipment. These costs can vary depending on factors such as the type of road, terrain, and location. It is crucial to conduct a comprehensive analysis of these costs to estimate the financial resources required for road network planning accurately.

The cost of land acquisition is a significant consideration in road construction projects. Acquiring land for road expansion or new road construction can be a complex and costly process. It involves negotiations, compensation, and potential legal challenges. Proper assessment and budgeting for land acquisition costs are crucial to avoid delays and cost overruns in road network planning.

Materials, such as asphalt, concrete, and steel, also contribute to the overall cost of road construction. The availability and cost of these materials can vary, impacting the budgeting and planning process. Additionally, labor costs, including wages and benefits for construction workers, machinery operators, and engineers, need to be factored in when estimating the cost of road projects.

Maintenance costs are another significant aspect of road network planning. Regular maintenance activities, such as repairs, resurfacing, and signage, are necessary to ensure the road network's longevity and safety. Allocating sufficient funds for ongoing maintenance is crucial to prevent deterioration and costly repairs in the future (Pricing European Transport Systems).

2.2. Impact of Cost on Feasibility and Sustainability

The cost of road construction and maintenance directly impacts the feasibility and sustainability of road network planning in Bulgaria. High construction costs can limit the scope and scale of road projects, potentially leading to incomplete or delayed infrastructure development. It is essential to strike a balance between the desired road network expansion and the available financial resources.

Moreover, the sustainability of the road network relies on the ability to allocate adequate funds for ongoing maintenance and repairs. Neglecting maintenance due to budget constraints can result in deteriorating road conditions, increased accidents, and higher long-term costs. Therefore, it is crucial to consider the lifecycle costs of road projects and allocate sufficient funds for both construction and maintenance activities.

2.3. Need for a methodology for cost-benefit analysis of infrastructure objects

The need for a methodology for cost-benefit analysis of infrastructure objects is related to the fact that such projects are usually large and long-term investments that have a significant economic and social effect on society. A well-developed cost-benefit analysis methodology can help assess the economic viability of infrastructure projects and make informed decisions about their financing and implementation.

Such a methodology usually includes the following stages:

- Identification of benefits and costs - this involves determining all the benefits and costs associated with the infrastructure project. This may include financial benefits such as profit from operating the site, but also non-financial benefits such as improved service quality or reduced pollution.

IOP Conf. Series: Materials Science and Engineering 1297 (2023) 012017 doi:10.1088/1757-899X/1297/1/012017

 Valuation of benefits and costs - once the benefits and costs have been identified, they must be valued financially or by other measures. This may involve using methods such as net present value (NPV), internal rate of return (IRR) or socio-economic analysis.
 Comparison of benefits and costs - once the benefits and costs have been assessed, a comparison must be made between them. This may include the calculation of various metrics such as NPV or social value of investment (SROI). The goal is to determine whether the benefits outweigh the costs and whether the project is economically viable.
 Risk management - the cost-benefit analysis methodology should also include risk assessment and management. This includes identifying the potential risks and opportunities associated with the project and using methods to assess and manage those risks.

The methodology of cost-benefit analysis of infrastructure objects is necessary to assess the economic viability of such projects and to make informed decisions about their financing and implementation. It helps identify and evaluate benefits and costs, compare them, and manage risk.

2.4. Return in investment in road infrastructure projects

The return on investment (ROI) in infrastructure projects can have significant benefits for a country. Here are some key points regarding the return on investment in infrastructure projects:

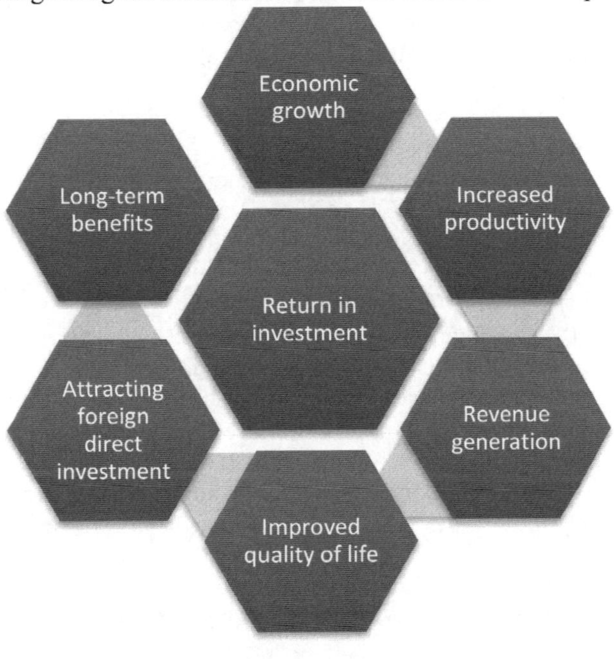

Figure 1 Return in investment

Infrastructure investments, such as building roads, bridges, airports, and ports, can stimulate economic growth. These projects create jobs, attract businesses, and improve transportation and logistics, leading to increased productivity and economic activity.

Infrastructure projects can enhance productivity by reducing transportation costs, improving connectivity, and facilitating the movement of goods and services. This can result in increased efficiency and competitiveness for businesses, leading to higher returns on investment.

Infrastructure projects can generate revenue for the government through tolls, fees, or taxes. For example, a new highway or bridge may generate revenue from toll collections, which can contribute to the return on investment over time.

Infrastructure investments in sectors like healthcare, education, and public utilities can improve the quality of life for citizens. Access to better healthcare facilities, schools, clean water, and sanitation can have long-term positive impacts on society, leading to improved social well-being and human development.

Well-developed infrastructure can attract foreign direct investment by providing a conducive environment for businesses. Investors are more likely to invest in countries with reliable transportation networks, efficient logistics, and modern utilities, which can contribute to the return on investment for the country (Technical Committee Evaluation, 2010).

Infrastructure projects often have long lifespans and can provide benefits for many years. While the initial investment may be significant, the long-term returns can outweigh the costs, especially when considering the economic growth, increased productivity, and improved quality of life that these projects can bring.

It is important to note that the return on investment in infrastructure projects can vary depending on various factors such as project planning, execution, maintenance, and the overall economic and social context. Proper evaluation, monitoring, and management of these projects are crucial to ensure optimal returns for the country.

Determining the return on investment in infrastructure projects requires gathering accurate and reliable data, as well as realistic forecasts of future revenues and expenses. This requires thorough research and analysis of the project, as well as considering all risks and uncertainties associated with the investment.

3. Expected traffic volume

The expected traffic volume is a crucial economic factor that significantly influences the planning activities of the road network in Bulgaria. It plays a vital role in determining the design, capacity, and efficiency of the road infrastructure.

3.1. Examination of the Expected Traffic Volume in Bulgaria

Accurately estimating the expected traffic volume is essential for effective road network planning. This involves analyzing current traffic patterns, population growth projections, economic development forecasts, and transportation demand. Data from traffic surveys, population censuses, and economic indicators can provide valuable insights into the expected traffic volume.

In Bulgaria, the expected traffic volume varies across different regions and road segments. Urban areas, major highways, and industrial zones generally experience higher traffic volumes compared to rural areas. Understanding these variations is crucial for prioritizing road projects and allocating resources effectively.

3.2. Influence of Traffic Volume on Road Network Design and Capacity

The expected traffic volume directly impacts the design and capacity of the road network. Roads with high traffic volumes require wider lanes, additional lanes, and advanced traffic management systems to ensure smooth traffic flow and reduce congestion. On the other hand, roads with lower traffic volumes may require less extensive infrastructure.

Accurate estimation of traffic volume helps determine the appropriate road design, including the number of lanes, intersections, and access points. It also influences decisions regarding the inclusion of features such as pedestrian walkways, bicycle lanes, and public transportation infrastructure.

3.3. Importance of Accommodating Future Traffic Growth

In addition to the current traffic volume, it is crucial to consider future traffic growth in road network planning. Population growth, economic development, and urbanization can lead to increased transportation demand over time. Failing to account for future traffic growth can result in inadequate road capacity, congestion, and reduced efficiency of the road network.

IOP Conf. Series: Materials Science and Engineering 1297 (2023) 012017 doi:10.1088/1757-899X/1297/1/012017

By considering future traffic growth, road planners can incorporate provisions for expansion, such as reserving land for future road widening or implementing intelligent transportation systems to optimize traffic flow. This proactive approach ensures that the road network remains efficient and can accommodate increased traffic demand in the long run.

Accurate estimation of expected traffic volume and consideration of future growth are essential for optimizing road network planning. By aligning road design and capacity with the expected traffic volume, Bulgaria can develop a robust and efficient road network that meets the transportation needs of its population and supports economic development.

4. Economic development and road network planning in Bulgaria

The relationship between economic development and road network planning in Bulgaria is significant. A well-planned and efficient road network plays a crucial role in supporting economic growth, regional development, and improving connectivity within the country. This section explores the interplay between economic development and road network planning in Bulgaria (OECD, 2021).

- Economic Growth and Regional Development: A well-connected road network is essential for fostering economic growth and regional development. It facilitates the movement of goods, services, and people, enabling businesses to access markets, suppliers, and customers more efficiently. By strategically planning road infrastructure, Bulgaria can promote economic activities in different regions, attract investments, and create employment opportunities.
- Trade and Commerce: A robust road network is vital for facilitating trade and commerce within Bulgaria and with neighboring countries. Efficient transportation links enable the movement of goods to and from industrial zones, ports, and logistics hubs. This connectivity enhances Bulgaria's competitiveness in the global market, attracts foreign direct investment, and promotes international trade.
- Tourism and Hospitality: Bulgaria's road network plays a crucial role in supporting the tourism and hospitality industry. Well-maintained roads provide access to popular tourist destinations, historical sites, and natural attractions. By improving road connectivity to these areas, Bulgaria can attract more tourists, boost the hospitality sector, and generate revenue for local communities.
- Accessibility and Connectivity: A well-planned road network improves accessibility and connectivity, particularly in rural and remote areas. It reduces travel times, enhances mobility, and connects communities to essential services such as healthcare, education, and employment opportunities. This inclusivity contributes to reducing regional disparities and improving the overall quality of life for Bulgarian citizens.
- Industrial and Logistics Development: The availability of a well-connected road network is crucial for industrial and logistics development. It enables the efficient movement of raw materials, finished products, and supplies, supporting manufacturing and distribution activities. By strategically planning road infrastructure near industrial zones and logistics centers, Bulgaria can attract businesses, stimulate economic growth, and create a favorable business environment.
- Infrastructure Investment and Job Creation: Road network planning involves significant infrastructure investment, which stimulates economic activity and job creation. Construction and maintenance projects generate employment opportunities in various sectors, including engineering, construction, and transportation. This investment not only improves the road network but also contributes to the overall economic development of Bulgaria.
- Sustainable Development: Road network planning in Bulgaria also aligns with sustainable development goals. By promoting efficient transportation systems, reducing congestion, and integrating sustainable practices, such as promoting public transportation and encouraging cycling and walking, road network planning contributes to environmental sustainability and reduces carbon emissions.

In conclusion, economic development and road network planning in Bulgaria are closely intertwined. A well-planned and efficient road network supports economic growth, regional development, trade, tourism, and connectivity. By strategically investing in road infrastructure, Bulgaria can enhance its competitiveness, attract investments, create employment opportunities, and improve the overall quality of life for its citizens. Moreover, integrating sustainable practices in road network planning ensures a greener and more sustainable future for the country.

5. Challenges in Bulgaria's economy and road network planning

5.1. Economic Challenges

- Low GDP per capita: Bulgaria has one of the lowest GDP per capita among European Union member states. This limits the available financial resources for road network planning and infrastructure development.
- Regional Disparities: There are significant regional disparities in Bulgaria, with some areas experiencing higher levels of economic development than others. This poses challenges in allocating resources and prioritizing road network planning projects to address regional imbalances.
- Brain Drain: Bulgaria has been facing a brain drain, with many skilled professionals leaving the country in search of better economic opportunities abroad. This can impact the availability of skilled labor for road network planning and implementation (World Bank, 2017).

5.2. Strategies to Overcome Economic Obstacles in Road Network Planning

- Diversifying Funding Sources: Bulgaria can explore alternative funding sources, such as public-private partnerships, international loans, or grants, to supplement limited government budgets. This can help attract private investment and leverage external funding opportunities.
- Prioritization and Efficiency: Prioritizing Road network planning projects based on their economic and social benefits can ensure efficient use of limited resources. Conducting cost-benefit analyses and focusing on projects that have the highest impact can help overcome economic obstacles.
- Enhancing Regional Development: Addressing regional disparities through targeted road network planning can help stimulate economic growth in underdeveloped areas. This can involve improving connectivity to attract investments, promote tourism, and create employment opportunities.
- Retaining Skilled Professionals: Implementing measures to retain skilled professionals, such as providing competitive salaries, career development opportunities, and a favorable work environment, can help overcome the brain drain and ensure a skilled workforce for road network planning activities.

In conclusion, Bulgaria's economic challenges, including low GDP per capita, regional disparities, and brain drain, have implications for road network planning activities. Limited funding, inadequate infrastructure, and skill shortages can hinder the development and maintenance of the road network. However, strategies such as diversifying funding sources, prioritization, regional development, and retaining skilled professionals can help overcome these obstacles and promote sustainable road network planning in Bulgaria.

6. Opportunities for economic growth through road network planning

Road network planning plays a crucial role in driving economic growth by improving connectivity, facilitating trade, and attracting investments. This section highlights potential economic benefits resulting from road network planning and the role of road networks in enhancing trade opportunities and attracting investments.

6.1. Economic Benefits of Road Network Planning

Figure 2 Economic benefits of road network planning

Well-planned road networks improve connectivity between regions, cities, and economic centers. This enables efficient movement of goods, services, and people, reducing transportation costs and enhancing economic integration.

Improved road networks facilitate the movement of goods, enabling businesses to access larger markets and expand their customer base. This can lead to increased trade activities, both domestically and internationally.

Road network planning and construction projects create employment opportunities, both directly and indirectly. The construction sector benefits from job creation during the road infrastructure development phase, while improved connectivity can attract businesses and industries, leading to additional job opportunities.

Well-planned road networks can stimulate regional development by connecting previously isolated or underdeveloped areas. This can attract investments, promote tourism, and encourage economic activities in these regions (Грозданов, 2019).

6.2. Role of Road Networks in Attracting Investments:

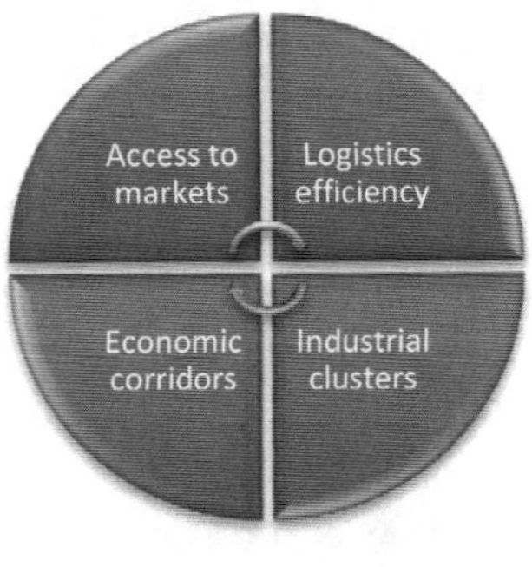

Figure 3 Attracting investments

Road networks provide businesses with access to markets, enabling them to reach customers and suppliers more efficiently. This accessibility can attract domestic and foreign investments, as companies seek locations with good transportation infrastructure.

Efficient road networks reduce transportation costs and improve logistics operations. This can make a location more attractive for businesses, as it allows for streamlined supply chains, faster delivery times, and cost savings.

Well-planned road networks can also support the development of industrial clusters by providing transportation links between manufacturing facilities, suppliers, and distribution centers. This clustering effect can attract investments and foster collaboration among businesses.

Road networks can create economic corridors that connect different regions or countries, promoting trade and investment along these routes. These corridors can attract infrastructure development, logistics hubs, and other economic activities.

Road network planning offers significant opportunities for economic growth. Enhanced connectivity, increased trade, job creation, and regional development are among the potential economic benefits. Road

IOP Conf. Series: Materials Science and Engineering 1297 (2023) 012017 doi:10.1088/1757-899X/1297/1/012017

networks also play a crucial role in attracting investments by providing access to markets, improving logistics efficiency, supporting industrial clusters, and creating economic corridors. Furthermore, well-planned road networks improve trade opportunities by facilitating cross-border trade, enhancing export competitiveness, and promoting tourism development (Европейска комисия, 2008).

7. Integration of sustainable practices in road network planning

The importance of Considering Sustainability in Economic Decision-Making can be generalized in:

- Long-Term Perspective: Considering sustainability in economic decision-making ensures that the impacts of road network planning are assessed not only in the short term but also in the long term. This helps avoid negative consequences and promotes the well-being of future generations.
- Environmental Protection: Sustainable practices in road network planning minimize the negative environmental impacts, such as air and noise pollution, habitat destruction, and carbon emissions. This is crucial for preserving natural resources and mitigating climate change.
- Social Responsibility: Integrating sustainability in economic decision-making demonstrates social responsibility by considering the well-being of communities, including their health, safety, and quality of life. It promotes equitable access to transportation and minimizes social disruptions caused by road infrastructure projects.

Benefits of Sustainable Road Network Planning for the Economy and the Environment are sustainable road network planning can lead to cost savings in the long run. For example, incorporating energy-efficient lighting systems or using recycled materials can reduce operational and maintenance costs. Additionally, sustainable road networks can attract investments, promote tourism, and enhance economic development. Sustainable practices in road network planning contribute to environmental protection by reducing carbon emissions, preserving natural habitats, and improving air and water quality. This helps mitigate climate change, protect biodiversity, and enhance overall environmental health.

Integrating sustainable practices in road network planning is essential for considering the long-term impacts on the economy, environment, and society. Bulgaria has been incorporating green infrastructure, sustainable materials, and active transportation elements in its road network planning. The benefits of sustainable road network planning include economic efficiency, environmental protection, improved health and well-being, and enhanced resilience. By prioritizing sustainability, Bulgaria can create a road infrastructure that supports economic growth, protects the environment, and promotes the well-being of its citizens.

By implementing a sustainable targeted policy for planning the development of the national road network in the long term, about 20% of the total budget can be saved, this can be generally summarized in the number of population benefited by the projects, net added value according to the standard of purchasing power , net jobs created and the decision-making sequence.

8. Conclusion

The importance of considering economic factors as a key component in the planning activities of the national road network emphasizes that economic considerations should be integrated into decision-making processes to ensure the efficient allocation of resources and maximize the benefits derived from road infrastructure investments.

Economic factors play a crucial role in determining the feasibility and viability of road network planning activities. It emphasizes the need to assess the economic impact of road projects, including their potential to stimulate economic growth, create employment opportunities, and attract investments. By considering economic factors, decision-makers can prioritize projects that have the potential to generate significant economic benefits for the country.

Furthermore, it emphasizes the importance of cost-benefit analysis in evaluating road network planning activities. It suggests that a comprehensive assessment of the economic costs and benefits associated with road projects is essential to make informed decisions. This analysis should consider factors such as construction costs, maintenance expenses, travel time savings, and the potential for increased economic activity.

The article also highlights the need for collaboration between various stakeholders, including government agencies, private sector entities, and local communities, to ensure that economic considerations are adequately addressed in road network planning. It emphasizes the importance of engaging with experts in economics and transportation planning to conduct thorough analyses and provide valuable insights.

Overall, the article underscores the significance of economic factors in the planning activities of the national road network. It emphasizes the need for a comprehensive understanding of the economic impact of road projects and the importance of integrating economic considerations into decision-making processes. By doing so, Bulgaria can optimize our road infrastructure investments and contribute to sustainable economic development.

As a result of the analysis of the impact of transport on the social aspect of the sustainable development of the country's economy, recommendations can be made in the following directions: - Ensuring access to goods and services for all members of society requires requirements for a higher degree of security and transport safety and higher comfort. On the other hand, the increase in the volume of transport and the increase in environmental risk lead to the opposite effect. Improving the quality of transport services, including personal safety, reducing the number of transport accidents and the risk to people's health, protecting the rights of passengers and access to transport for the population from remote and peripheral areas are extremely important and should be prioritized in transport policy, as they form significant external costs borne by the whole society. The need for socially sustainable transport is related to the provision of safe, accessible and efficient transport services and conditions for movement by personal motor vehicles. In addition, society wants the country's transport policy to ensure the safety of human life, create prerequisites for the creation of jobs and ensure a reduction of the harmful impact on the environment. To achieve these goals, actions are needed to increase road safety. In addition to the purely human factor, there is also a strong economic logic for investments in road safety – studies conducted at the EU level show that for 1 euro of investment in road infrastructure, around 3 to 14 euro of savings are expected from fewer casualties, - low health care costs and damage, as a result of reducing the number of road accidents. All this should be taken into account when evaluating infrastructure projects together with the effect of reducing travel times, the benefits of improved road infrastructure and the reduction of transport costs for society as a whole.

References

[1]Technical Committee Evaluation. (2010). *Financing of Road Infrastructures: Guide for New Methods of Financing and Public/Private Partnership.*

[2]Calculating Transport Congestion and Scarcity Costs. (2008). *Final Report of the Expert Advisors to the High Level Group on Infrastructure Charging.*

[3]European commission. (2022). *https://europa.eu/next-generation-eu/index_bg.* https://europa.eu.

[4]OECD. (2021). *COVID-19 emergency government support and ensuring a level playing field on the road to recovery.*

[5]Pricing European Transport Systems. (н.д.).

Tsigdinos, S. (2020). *Strategic road network formulation: developing an alternative methodology towards sustainable mobility.* Transportation Research Procedia.

[6]World Bank. (2017). *GLOBAL MOBILITY REPORT* . Washington: WORLD BANK.

[7]Grozdanov, H. (2019). Impact of external costs from road accidents on the economy of Bulgaria in 2019. Sofia: XIII National Conference on Transport Infrastructure with International Participation, October 2020, pp. 82 to 89.

[8] European commission. (2022). https://europa.eu/next-generation-eu/index_bg. Retrieved from https://europa.eu.european transport safety council. (2021, August 1st). etsc.eu/eu-road-

IOP Conf. Series: Materials Science and Engineering 1297 (2023) 012017 doi:10.1088/1757-899X/1297/1/012017

deaths-down-by-3900-in-2020/. Retrieved from etsc.eu: https://etsc.eu/eu-road-deaths-down-by-3900-in-2020/
[9]OECD. (2021). oecd.org. Retrieved from https://www.oecd-ilibrary.org/?_ga=2.188167914.397094570.1627217129-1162414499.1627217129. Pricing European Transport Systems.
[10] Ministry of Transport and Communications. (2017). Integrated transport strategy in the period until 2030. Sofia: MTC.

IOP Conf. Series: Materials Science and Engineering 1297 (2023) 012018 doi:10.1088/1757-899X/1297/1/012018

Possibilities for the construction of a parallel route of the European transport corridor No 9 through Romania, Bulgaria, and Greece

Nikolay Tsonkov[1,3], Kamen Petrov[2]

[1] Regional Development Department, University of National and World Economy, Sofia, 1517, Bulgaria

[2] Regional Development Department, University of National and World Economy, Sofia, 1000, Bulgaria

[3] kolio_tzonkov@abv.bg

Abstract. The Bulgarian state is significantly behind in the development of its strategic road infrastructure. This circumstance is complemented by the deteriorated state of the already-built infrastructure in the Bulgarian state. There is a need to change the policy philosophy for managing and maintaining the strategic infrastructure in Bulgaria. It is necessary to strongly link this policy with the regional development policy, due to the similar objectives they must achieve. They are aimed at bringing Bulgarian regions closer together, and their economic and regional development based on improved connectivity between them. Achieving all this, however, is hampered by a significant deficit in the system. There is a lack of connectivity and development of the transport axes in the North-South direction. In this regard, the authors set as their main objective to analyze and assess the possibilities for the construction of a parallel route of the European transport corridor No. 9 through Romania, Bulgaria, and Greece, which includes the construction of a tunnel between the town of Troyan and the village of Karnare in the central part of the Balkan Mountains. The implementation of this project will bring economic, social, transport, and financial benefits to the Bulgarian state. The construction of the project ensures connectivity between the Balkan countries, the formation of a single European market, stimulating the development of Northern Bulgaria, and a targeted impact on the North and South-Central planning regions.

1. Introduction

A transport corridor is a set of high-tech technical links and capacities enabling transport and transmission in a particular direction. Due to the possibility of transport links to create spatial and functional structures, often affecting the form of a network of urban centers or the distribution of green spaces, their functioning has become the subject of planning undertaken at various levels, including the European Forum. Going back in time, the European Commission Report published in 1998 stressed that the role of the Pan-European transport corridors was an essential element and a key factor in increasing the competitiveness of European communities ensuring sustainable jobs, and strengthening socio-economic cohesion throughout the Union by ensuring effective communication with peripheral regions [1]. In 2014 EU updated these corridors and developed the Trans-European Transport Network (TEN-T) which was synchronized according to the sustainable development and mobility concept [2].

The strategic importance of transport corridors for the spatial development of Europe, especially in the context of their impact on the organization of space, has been assessed in official documents formulating the principles of the spatial development of the European continent, the territory of the

IOP Conf. Series: Materials Science and Engineering 1297 (2023) 012018 doi:10.1088/1757-899X/1297/1/012018

European Union and the Baltic Sea region as an area of cooperation between EU Member States and national institutions in Central and Eastern Europe in the field of spatial planning [3]. In practice, in more recent times, it can be assumed that the trans-European transport network is the starting point for the plan to consolidate the transport infrastructure network across the European continent. The plan, formulated in 1997, envisages the creation of 10 Pan-European Transport Corridors and Pan-European Transport Zones on the border of the territory of the European Union and the neighboring countries to the south and east.

In this presentation we will focus on transport corridor No. 9 and the possibility of building a branch of the corridor. We analyze the new transport branch in the context of the TEN-T connectivity development. In this sense, we consider the new corridor as a part of the Orient/Mediterranean route which passes through Bulgarian country. The need for such a proposal is rooted in the complex international economic and political relations in the twenty-first century which influence the development of South-Eastern Europe. In addition, the processes of regional development and the formation of connectivity between individual spaces and their adjacent regions have a significant impact on the development of the countries in the region [4;5]. Thus, in this paper, we will present the possibility of building an alternative branch of the European Transport Corridor No. 9 along the route Bucharest - Craiova- Turnu Magurele- Nikopol-Pleven-Lovech- Troyan- Karnare- Pazardzhik- Dospat-Gotse Delchev- Promakhon- Kavala-Alexandroupolis. This Pan-European corridor is connected to the Orient/Mediterranean European route according to the TEN-T network. The purpose of the construction of this alternative is to show that it contributes to the development of transport infrastructure, economic development, and free movement of people and products in Southeast Europe between the countries of Romania, Bulgaria, and Greece. This alternative route can overcome significant this stage at infrastructure barriers, which can be defined as a disadvantage and an obstacle to creating opportunities for economic growth in the mentioned regions where the alternative route passes [6;7]. For this reason, we accept that infrastructure investment needs to anticipate a need, which is reflected in planning requirements for assumptions about infrastructure expansion and upgrading at agglomeration, regional, and local levels. This also implies a new approach to regional development policies and, above all, a new approach to European policy planning and programming [8]. Without a long-term vision in planning the development of transport routes, we will face several barriers to economic growth in the regions and, in practice, a reduction in the quality of life of the population. It is worth stressing here that the infrastructural development of the transport system is a complex process that has specific problems that need to be addressed. This means an integrated approach to those territories that have a specific arrangement of technical (sectoral, functional), organizational, financial, and regulatory subsystems [9;10].

In today's world, regional development with its functional specificities is called upon to implement specific policies whose main skeleton is the infrastructure system that determines the transport accessibility to the individual territory and space [11]. The operational effects of this system are determined by the availability and quality of regional development policies and the transport services provided. The factor that is of major importance is transport, which is fundamentally a networked economic activity. The greater the usefulness of transport, the stronger it is and the more systemic the links between its various forms, in effect between the urban transport system and the interregional transport system, which has an impact on the proper functioning of the entire network. Urban transport infrastructure consists of the infrastructure of different transport sectors, which must be adapted to the specific characteristics of a given territorial area and at the same time linked to the transport networks of the region and the country. From a spatial planning point of view, the transport corridor development policy considers two possible solutions [12]. Firstly, densification of the network of connections with high technical parameters (motorways and main railways), which, however, has a short-term effect. Second, is the framing of infrastructure investment within broader spatial development programs, where less developed areas will benefit more from access to transport networks [13].

2. Methodological framework and some author's remarks

In the report, the authors mainly examine the trans-European transport corridors that are still in operation, but after 2014 are being upgraded and complemented by the TEN-T network developed and adopted by the EU. The authors' understanding is that the trans-European transport corridors provide the main connectivity between countries based on the construction of new road routes. At the same time, the developed TEN-T network should be seen in the context of the sustainable development of European regions and building sustainable mobility. The TEN-T network is therefore based on the development of rail transport because it is the most environmentally friendly, safe, fast, and can provide a high degree of mobility. A careful review of the TEN-T network shows that only two corridors pass through Bulgaria. These are the Rhine-Danube and the Orient-Mediterranean corridor. The economic and regional development of the Bulgarian state implies the creation of transport connectivity in the Balkans and the construction of common infrastructure projects to form a common Balkan market. The new corridors cannot fulfill this objective. It is therefore necessary to first build trans-European transport corridors through Bulgaria, which should be oriented North-South. In this sense, the main objective of the authors is to analyze and assess the possibility of building a branch of corridor number 9 (Figure 1), which overlaps to some extent with the new route of the Orient/Mediterranean corridor. However, there is a North-South connectivity deficit. The construction of new bridge structures on the Danube River in the Bulgarian territory of the river can be seen in the context of the trans-European transport corridors, which can be modified and oriented North-South. And this is a significant deficit of the EU in terms of connectivity of European regions. To realize the formulated purpose the authors, use geographical, territorial, and networking approaches. The authors mainly use analytical tools, spatial and descriptive analysis, forecasts, and assessment tools.

3. Direction of development of the European transport corridors and regional development policies

The aim of developing the network of corridors and their branches is to ensure the territorial cohesion of the EU and to facilitate the free movement of people and goods. In addition, the focus of the EU Transport Strategy is to create integrated, low-emission transport in urban agglomerations, concentrated in a network of lines of individual means of transport and with a common fare system. The urban transport system in its organization, including the timetable, must be adapted to the interregional system. For the transport system to function optimally, it needs solutions to improve connectivity. From this point of view, an important element for improving the transport system is the development of transport corridors and the transport infrastructure included in them. The lack of bypasses for both rail and road transport contributes to communication problems in the development of entire regions and their adjacent villages and towns. In addition, we can consider that the different forms of transport in logistics chains occupy about 80% of all activities, and therefore have a significant impact on the cost and quality of logistics services. In the transport policy adopted, the European Union aims to combine the main modes of transport, i.e., rail, road, sea and inland waterway transport, and air transport, into a flexible transport network, using their best characteristics most efficiently, especially by transporting goods in a single freight unit [14].

The corridors are designed to coordinate different projects at the transnational level. They should contribute to the development of the infrastructure of the core network to address bottlenecks, increase cross-border connectivity, and improve the efficiency and sustainability of the transport system. They should also contribute to improving the cohesion of EU regions through better territorial cooperation [15]. According to the EU's intentions, the core network corridors are to be implemented by 2030. In this direction, building transport infrastructure linking countries is of strategic importance for the development of the economy both continentally and globally. The removal of restrictions on the movement of goods from countries that are members of the European Union is one of the key elements that has had a positive impact on trade. Moreover, the intensification of goods flow development and appropriate upgrading of linear and nodal infrastructure has an impact on economic progress and outreach.

4. Development of regional logistics services

The trans-European transport network TEN-T includes road, rail, air, sea, and river routes, which are the most important links in terms of the development of the European Union, as well as point infrastructure elements in the form of sea, airport, inland ports and roads and rail terminals [16]. In this direction, the branching of transport corridors may also require the construction of intelligent systems and new regional hubs to further develop individual regions [17]. In addition, intelligent transport systems should be an integral component of these routes, the deployment of which contributes to improving network capacity, traffic safety, and reducing transport pollution. Over the last ten years, the global economy has seen dynamic progress in new technologies.

The TEN-T network consists of a core and a wide periphery. The core of the network is to be built by 2030 and the rest by 2050. Thus, the concept of sustainable development of regions through rail and other transport mobility is adopted. The main connectivity is achieved through the construction of railway infrastructure. The implementation of the new concept will help to stimulate the economic development of the EU and the regions through which the routes pass [18]. In line with the new mobility concept, 9 corridors have been developed (Figure 2), two of which pass through Bulgaria. These are the Rhine-Danube and the Orient/Mediterranean. The routes through the Bulgarian state cover the following routes:

- Orient/Mediterranean with the following routes: Hamburg - Berlin; Rostock - Berlin - Dresden; Bremerhaven/Wilhelmshaven - Magdeburg - Dresden; Dresden - Usti nad Labem - Melnik/Prague - Lysa nad Labem/Porzicani - Colin; Colin - Pardubice - Brno - Vienna/Bratislava - Budapest - Arad - Timisoara - Craiova - Calafat - Vidin - Sofia; Sofia - Serbia/North Macedonia border; Sofia - Plovdiv - Burgas/Turkey border; Turkey border - Alexandroupoli - Kavala - Thessaloniki - Ioannina - Kakavia/Igoumenitsa; North Macedonia border - Thessaloniki; Sofia - Thessaloniki - Athens - Piraeus/Ikonio - Heraklion - Lemesos (Vasiliko) - Nicosia/Larnaca; Athens - Patras/Igoumenitsa.
- Rhine-Danube with directions: Strasbourg - Stuttgart - Munich - Wels/Linz; Strasbourg - Mannheim - Frankfurt (Main) - Würzburg - Nuremberg - Regensburg - Passau - Wels/Linz; Munich/Nuremberg - Prague - Ostrava/Prscherov - Žilina - Košice - border with Ukraine; Wels/Linz - Vienna - Bratislava - Budapest - Vukovar; Vienna/Bratislava - Budapest - Arad - Moravica/Brasov/Krajova - Bucharest - Giurgiu/Constanta - Sulina.

The corridors thus described build on and complement the Cretan ones, which means that it is of particular importance for the Bulgarian state to build the routes along both concepts. The TEN-T network fully corresponds to the policy adopted by the Bulgarian government to modernize the railway routes to increase the security of supply and transport mobility. The Orient/Mediterranean Corridor passing through Bulgaria provides transport connectivity between Asia and Europe. The TEN-T network can thus be seen in the context of China's One Belt, One Road concept [19;20].

In the present study, the authors focus on the idea and the possibilities of building a route connecting Bulgaria and Romania, i.e. oriented in the North-South direction. Thus, such an infrastructural object should be analyzed in the context of the realization of the Orient/Mediterranean corridor [21]. At the same time, it is a new branch of corridor number 9 according to the concept of pan-European transport corridors. Another important aspect that should not be overlooked is the development of international economic and social relations, technical and technological progress, and rapid urbanization generating many global civilization problems. Thus, regional development is often defined as problematic due to the different speeds of development of different territories [22]. This is because large corporations very often transfer part of their production to regions where production costs are low, but others remain without any investment. However, local investment leads to an increase in transport demand. New flows of freight are also emerging and growing, and regional development is being modeled, necessitating new transport and logistics infrastructure. This means that an efficiently functioning transport system

IOP Conf. Series: Materials Science and Engineering 1297 (2023) 012018 doi:10.1088/1757-899X/1297/1/012018

within the European Union, and in particular the countries of South-Eastern Europe, must contribute to improving the functioning of the single internal market, stimulate the economic growth of the region, and increase the competitiveness of individual Member States and the EU vis-à-vis the rest of the world. This means that we need to be clear about the need to have a major impact on shaping freight flows and, therefore, freight transport markets. In practice, this means developing all modes of transport infrastructure, which also means efficient access to it, its density, and its technical condition. To exploit the potential of Bulgaria's strategic geographic location, it is necessary to ensure the appropriate quality not only of the road but also of the main rail corridors with a high-capacity logistics infrastructure [23]. Huge investments are required not only in the modernization of Bulgaria's linear transport infrastructure but also in the improvement of quality and productivity, as well as the creation of new infrastructure facilities for logistics services, including intermodal transshipment terminals and logistics hubs. In Bulgaria, it is necessary to accept that it is good to build modern line and point infrastructure in the form of intermodal terminals and logistics centers. Important for the efficiency of the system is the reliability of the means of transport and proper organization. An important element in the organization of transport processes is the creation of contractual transport routing channels in which large flows of freight will move along defined routes that are alternative but also high-speed. Organized and at the same time channelized flows significantly influence the proper planning of the development of transport infrastructure for both freight and passenger transport in the network and spatial system, both agglomeration and interregional.

The development of transport infrastructure has a significant impact on the economy and quality of life of regions and populations. In this direction, the construction of a new branch of Corridor No 9 in the direction Bucharest - Turnu Magurele - Pleven - Troyan - Karnare - Pazardzhik - Gotse Delchev - Promakhon - Kavala - Alexandroupolis will act as an effective link in the North-South direction with many geo-economic benefits.

First and foremost, stimulating the development of the region through joint investments and common cross-border projects, leading to regional development and the development of the supply of transport services ahead of current needs.

Secondly, to impose a model of spatial architecture, where the individual elements of the transport system, due to the wide range of their influence, are placed on an equal footing with spatial architectural and urban planning solutions and spatial plans. Next, creating the conditions for modern and sustainable environmental development of the regions.

5. Development of North-South transport connectivity on the new route

For several years now the issue of a complementary route has been very important because traffic on the Romania-Bulgaria-Greece (Turkey) route has increased 3-fold with increasing trends. Thus, the construction of this route means analyzing the need for transport services, the possibilities of securing funds and financing for investment and the actions to be taken, as well as the conditions to be met to facilitate access to such funds or financing and making appropriate recommendations. This means that the new corridor route must be taken seriously by the diplomacy of Bulgaria, Romania, Moldova, Hungary, and Greece. The route we are proposing should start from the southern part of the ring road around the city of Sofia. Bucharest and continue in the direction of Alexandria, and then in the direction of -Turnu Magurele- (bridge or ferry)- Nikopol- Pleven -Lovech - Troyan (tunnel) Karnare - Rozino - Hisarya, -Syedinenie- Pazardzhik-Peshtera-Dospat- Gotse Delchev-Ilinden-Drama- Kavala-Alexandroupolis.

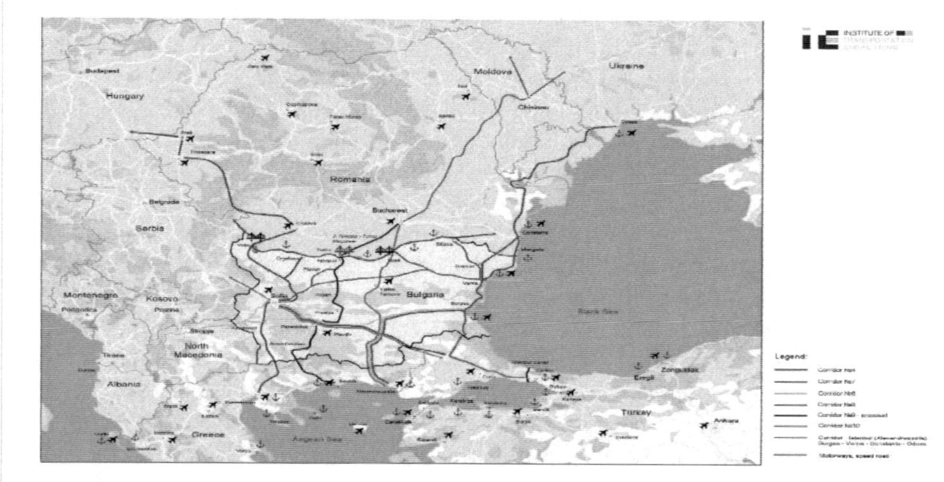

Figure 1. Alternative route (branch) of European Corridor No. 9.

Figure 2. TEN-T connection according to new regulation in 2014. Source: https://www.mtc.government.bg/bg/category/298.

In principle, this route could play the role of the crossing point of two of the ten pan-European corridors - Pan-European Corridor VII (Danube), which connects Western Europe with the Black Sea and Central Asia, and Pan-European Corridor IX, which connects Scandinavia and the Russian Federation with Southern Europe and Asia Minor, respectively. As a result of its geographical position, more than 80% of the transit of goods between Romania, Bulgaria, Greece, and Turkey, or about 23 billion euros per year, passes through Bucharest-Giurgiu-Ruse and Craiova-Beket-Orjáchovo. In this direction, the route is important as it passes through both Romania and Bulgaria. In Romania, the beginning of the branch leads from the southern transport arc. The southern ring road (A0) is part of the second ring (the ring), as it is known, which is 52 kilometers long. It is important for traffic because it is a second ring and supports the existing ring road, which can now function as an internal boulevard to collect and redistribute traffic within Bucharest. The new ring road will have a more regional and national function. It can be expected that the tandem of the two will provide long-term throughput capacity and, in effect, create the possibility for a new route along Corridor 9 in the direction of Alexandria and Turnu Magurele. The distance from Bucharest to Turnu Măgurele is 135 kilometers. It is important to note that Sector 01 - km 0+000 - km 66+000, Bucharest - Alexandria is foreseen to be part of the highway to the city of Craiova. At km 0+000, the first road junction is being made with Ring 2 of the Bucharest Ring Road, a ring that will be made in the profile of a motorway. The motorway alignment in this section runs almost parallel to DN6, on its left side, in a south-westerly direction, bypassing the towns of Mihailesti and Gornji. At km 13+200 of the motorway, respectively 29+500 of DN6, a road junction with DN6 is being built. On its right side, it bypasses the towns of Gimpati, Naipu, Bujoreni, Draganes-Vlaska, Komora, and Magura. At approximately km 23+000 of the highway (to the right approximately km 39+700 of DN6), the route crosses DN61 (km 4+000), near the town of Letca Noua. At km 66+000 the motorway route reaches the city of Alexandria, a city that will connect to the motorway, via a road junction. The city of Alexandria will be surrounded by the motorway from the north. To the south of Alexandria, an expressway should be designed from km 68+000 to km 135+000 to connect the highway to Turnu Magurele. The upgrading of the DN 52 road will become the main connecting artery between Alexandria County Municipality (Teleorman) and Turnu Măgurele Municipality, one of the major Romanian ports on the Danube. Here the Danube is a natural border between Bulgaria and Romania. Currently, the crossing points of the Danube between Romania and Bulgaria are the Giurgiu-Ruse (road and railway bridge), Calafat-Vidin (road and railway bridge), and by ferry - Calafat, Becket, Turnu Magurele, Giurgiu, Oltenica and Cularasi. The river separates the two countries by a great distance - 450 km, so there are not enough crossing points. In this direction, the town of Turnu Măgurele and the town of Nikopol are a good opportunity to build a new bridge across the Danube. In practice, the option of a bridge at Hiекopol - Turnu Măgurele provides the most direct connection between the capitals of Sofia and Букурекst. It will be conveniently located in the north-central part of the country. It will be relatively equidistant from the existing two bridges, providing a modern transport link with the new Hemus motorway, and will have a major effect on one of the poorest regions of the country. In this direction, there is a joint memorandum signed in October 2014, which foresees the construction of the bridge between Turnu Măgurele and Nikopol, a project that was supposed to be completed at the end of 2016, but there is no traffic on it until 2023. Of interest to both countries would be the restart of the project to build a hydro-technical junction connecting the cities of Nicopolis and Turnu Măgurele across the Danube. The Nikopol-Turnu Măgurele project envisages a 16-meter-wide barrier wall with a road and a railway between the Bulgarian and Romanian banks, and two hydroelectric power stations with a total capacity of 600 megawatts to be built on either side. The construction of the Nikopol-Turnu Magurele facility was designed in the 1960s and the first sod was laid for its construction in 1977. Its construction will ensure year-round river navigation, generate cheap and environmentally friendly energy, provide flood protection, sufficient water for irrigation, and the development of tourism and communications in the region.

Once the Corridor 9 route enters Bulgaria, the construction of an expressway with 4 lanes and a maximum speed of 120 kilometers in the direction of Nikopol - S. Debova - s. Koilovtsi - s. Grivitsa or 53,700 km of new road Beli Osam. This distance is about 78 kilometers, which can be divided into two lots of 40 kilometers each. Also, an expressway (4 lanes, 120 km speed limit) should be built to cross Hemus Motorway, around the village of Slatina. Thus, the route of Corridor No. 9 enters the most interesting section, because in addition to the construction of a road link between northern and southern Bulgaria, here thought about the construction of a railway link between Troyan and Karnare, which will connect with the Sub-Balkan railway. In terms of road connectivity, the distance from S. Beli Osam to. Karnare is 43 km, but if a tunnel is built under the Beklemeto pass with two pipes the distance can be reduced to between 18 and 25 km depending on where the tunnel is built. Back in 2002, 16 municipalities demanded that the government include in the national development plan the tunneling under the Balkan

IOP Conf. Series: Materials Science and Engineering 1297 (2023) 012018 doi:10.1088/1757-899X/1297/1/012018

Mountains - from Troyan to Karlovo. According to the mayors at the time, the tunnel would open the shortest route from the Danube to the White Sea and stimulate economic development on both sides of the Balkan. It is estimated that the route from Troyan to Karlovo will become 30 km, 12 of which will be underground. Ten years ago, on 20 November 2012, there was again a meeting of the regional governors of Lovech, Plovdiv, and Smolyan who talked about this project. Realistically speaking, the tunnel under Beklemeto will not be more expensive than 600-700 million euros and will pay for itself in 8-10 years given the traffic that is expected to be in the north-south direction. The same applies to the construction of a railway tunnel because it will make it possible for freight from Central Europe and Western Europe to reach Belomorje, Istanbul, and Asia quickly on the Craiova-Turnu Magurele-Nikopol-Pleven-Lovech-Troyan-Karnare railway and the sub-Balkan railway, which shortens the movement of train sets considerably and makes them cheaper and more efficient. In this direction, the Bulgarian state should seriously consider, together with Romania, the construction of a railway from Craiova-Turnu Magurele-Nikopol-Pleven-Troyan-Karnare-Plovdiv.

After the Troyan-Karnare corridor No. 9 has been passed, there is a possibility to continue southwards, where a new expressway (4 lanes, speed 120 km) could be built, starting around the village. Karnare towards the town of Hisarya, where a ring road should be built and continue along the Novo Zhelezare - Lyuben - Connection ring road. The route requires the construction of a cloverleaf on the Trakia highway - Dobrovnitsa - Eastern bypass of Pazardzhik - Kv. Glavinitsa - Peshtera.

This is practically a completely new route, which sets new opportunities for the development of the regions around which the route passes. This also means attracting new investments in these regions and settlements, and therefore a new role for the city of Pazardzhik as an intermediate center between Sofia and Plovdiv. Another alternative route within Corridor 9 is the construction of an expressway with 4 lanes and a permissible speed of 120 km in the direction from the village of Karnare to the town of Pazardzhik. Klisura - s. Slatina - village. Bogdan - village of. Kliment - s. Voinyagovo - s. Pesnopoy - s. Streltsi - gr. Brezovo - s. Brothers Daskalovi- s. Cherna Gora - connection with "Trakia" and "Maritsa" highway. In this way, the conditions for the fastest access to Istanbul and Asia from Central Europe will be created. This will also have a positive impact on settlements such as Brezovo and the municipality of Brat'a Daskalovi, as well as on the towns of Parvomay and Chirpan. This is also the place to draw attention to the importance of the Sofia-Karlovo-Kazanlк-Sliven-Vetren Road, which is high time was declared by the government to be an alternative to the Trakia motorway. This means that the Sofia-Karlovo-Sliven-Aytos-Karnobat route should be included as a priority and designed and widened to a 20 m expressway with four lanes and a speed of 120 km. It is important to note that work on the modernization of this road has been underway since 1968, but in practice, the necessary funding has not been found and the modernization of the Sofia-Sliven-Karnobat-Vetren road remains for better times to come.

Returning to the alternative route of Corridor No. 9 at Peshtera, it should be noted that this part is where the significant difficulties in its construction lie. It is necessary to work on the design of a bypass of the town of Peshtera and the town of Batak and then the continuation to Dospat. The distance from the town of Peshtera to the town of Dospat is 75.13 kilometers, and the terrain is quite difficult, for us, the real alternative is a three-lane road with a speed of 90 kilometers.

However, the possibility of designing an expressway with 4 lanes can be built in about 1/3 of the route, which we consider appropriate and possible, but for sure, will make construction more expensive. On the other hand, this is a justified expenditure, because it will have a positive effect on the development of tourism and the livelihood of the population of this mountainous part of Bulgaria. The situation is similar on the Dospat-Gotse Delchev road, where a road with 3 lanes and a speed of 90 kilometers must be designed, in the part with. Satovcha - Gotse Delchev with 29, 280 km. A 4-lane road with a speed of 120 km should be built. Shortly before the town of Gotse Delchev, the road diverges towards the Ilinden and Esokhi BCPs, crosses into Greek territory, and reaches the town of Drama. The distance between Gotse Delchev and Drama is 73,720 km, The road is first class, but in our opinion, there is a possibility to build an expressway to increase the permeability of the route and to make it faster in the north-south direction. The route then heads in the direction of Drama-Kavala at 136,740 km. This road could also be improved and a 4-lane expressway with a speed limit of 120 km could be built. Shortly before the town of Kavala, the route meets the Neia Igtania highway, which naturally connects it to Alexandroupolis.

Within the European economic area, these roads are of utmost importance, improving connectivity between the different territories and enabling regional development policies. In the twenty-first century, railways and motorways have become increasingly important, forming the backbone of Europe's transport system. In this direction, attempts are also being made to develop container terminals and logistics centers, which act as integration points for certain parts of the transport chain, through the European transport corridors. The areas in which they operate, however, are most often not only the logistics and regional infrastructure itself, but also places in settlements where special economic zones, industrial and technological parks, or free zones are important for the socio-economic development of the territories. Thus, the spatial planning of industrial zones should provide for future transport services by the recommendations of EU transport policy. The routes of the main and branch lines of the European transport corridors are important. We have not yet fully developed all the routes of the 5 corridors that pass through Bulgaria. In addition, we must construct the new two corridors according to the TEN-T connectivity. More persistence and courage are needed on our part to be able to implement these large-scale projects, which are important not only for Europe and Asia. It should be noted that the intermodal transport market is a transport market with great opportunities. Growing transport needs, together with economic progress, are leading to dynamic regional development and attracting new investments. An opportunity to increase dynamism and accelerate development may result from the modernization of existing and the construction of new infrastructure, transport, and logistics centers located outside urban agglomerations. Harmonization of interventions between cross-border levels - national-regional-urban - to prevent conflicts between the objectives and outcomes of projects at different levels (e.g., suburban road projects undermining the transition to sustainable urban mobility). All this is linked to the goal of modern regional development that is sustainable. This means that the construction of regional infrastructure must maximize the functional synergy between the various national and regional subsystems within the European Union. This will enable proper transport services, and regional connectivity, attract investment, and at the same time manage the flow of freight, and free movement of people and services.

6. Conclusion

The construction or modernization of new routes along the European transport corridors is extremely important for the Bulgarian state (especially the Rein-Danube and Orient/Mediterranean corridors). This policy is complemented by the implementation of a targeted regional policy to ensure growth, quality of service, the possibility of storing intermodal units, and expanding the scope of logistics services at a regional level. These are the elements that can frame new regional development policies. The new Corridor 9 route is an opportunity for Bulgaria to create the conditions for several regions to develop, providing the framework in which to do so. It is also an opportunity for the country's businesses to get a boost. The new route must be built by the Orient/Mediterranean corridor which passes through Bulgaria. In this sense, it is important to modernize the regions by improving and building new rail and road links, providing a new type of freight transport, such as containers, which will bring a new dynamic to all modes of transport. In practice, the branching of the route from Bucharest via Pleven, Gotse Delchev, Kavala, and Alexandroupolis will strengthen the European Union's transport policy, the convergence of the regions, and thus support the balanced economic development of the various transport sectors and regional development between Bulgaria, Greece, and Romania. In practice, the free movement of goods and services will strengthen the possibility of forming a single European market. Its creation is not possible without adequate, systematic logistical support from Bulgaria, Romania, and Greece for regional development. Thus, the development of international transport corridors and the maximum use of transit potential meet the strategic interests of the Russian Federation. The geopolitical location allows Bulgaria to receive significant revenues from the export of transport services, including the implementation of transit flows. Bulgaria can take a significant position in the global economy as a transport bridge between Europe and Asia (in the West-East and north-south directions), provided that the transport system does not hinder but stimulates the sustainable development of the Bulgarian economy. The qualitative state of Bulgaria's transport infrastructure still creates significant constraints to the optimal use of transport potential. In practice, it is not an element conducive to the development of the economy and the increase of foreign trade. In this direction, we should also strive with all measures and policies to build the regional and national infrastructure. The routes of Corridor No. 9 play an important role in the transport policy of Bulgaria in the context of emerging problems of creating an active innovative strategy for the development of the transport system of the country. In this regard corridor No. 9 mitigates the effects of technical and technological backwardness. The necessary capital participation of the state grows because of the adaptation of the infrastructure towards international standards. These rules of the game require the construction of service infrastructure complementing transport corridors and the construction of additional alternative routes for the regional development of the country. It must also be reflected in an adequate attitude to the management of capital assets, achieved through projects co-financed by the European Union (to prevent the premature degradation of infrastructure because of failure to carry out appropriate maintenance/upgrading). Continuous review and flexibility of design principles and standards as well as those related to functions is needed. This will reduce the investment costs associated with projects and bring standards in line with current technological realities. Bulgaria needs to have its commitments on infrastructure maintenance - funding requirements, maintenance organization/contracting practices, etc. All these actions will have concrete and immediate benefits in terms of the development and specialization of the workforce in direct relation to the construction market, including economic agents, people managing infrastructure projects and roads, and transport operators. It must be stressed that to have effects and results in the current and future financial period, all these actions with the development of European transport corridors and the policies for their implementation need to be undertaken as soon as possible, bearing in mind that they are long-term measures that can be implemented during economic cycles of socio-economic development.

Acknowledgment
The paper is financed by a scientific project of UNWE number NI 5/ 2021.

References

[1] Implementation report on the TEN-T guidelines (Trans-European Networks - Transport) The time limit for receipt of tenders: 03/09/2004
[2] European Commission 2021 *Proposal for a Regulation of the European Parliament and of the Council on Union Guidelines for the Development of the Trans-European Transport Network*, Amending Regulation (EU) 2021/1153 and Regulation (EU) No 913/2010 and Repealing Regulation (EU) 1315/2013. Available online: https://eur-lex.europa.eu/legal-content/EN/ALL/?uri=COM%3A2021%3A812%3AFIN (accessed on 11 July 2022).
[3] Krustev K, Donchev D, Lozanov E, Karakashev Hr, Karastoyanov S, Dokova S *Geoeconomics* 2008, 273.
[4] Tonkova S, *Theoretical Models and Approaches for Implementation* 2002, 394.
[5] Stoencheva T 2010 *Regional Economics* (Sofia: Stopanstvo Publishing house) p 159.
[6] Koster R, Petrov K 2020 *Infrastructure Corridor 9 as a factor for improving transport accessibility in Bulgaria* (*Geopolitics* Vol. 2) p 74 - 85
[7] Grozdanov H 2021 *Need for connectivity between Bulgaria and Romania across the Danube (Proc. of XIVth National Conference on Transport Infrastructure with International Participation)* p 111-114.
[8] Shishmanova M 2013 *Regional development and spatial planning should be seen as a single process* (*Economics and Management* Vol. 4) p 52-69.
[9] Dimitrov S, Dimov N 2002 *Proceedings Int. Scientific conference in memory of Prof. D. Yaranov* (Varna: Varna Economic University publishing house) p 56 - 70

[10] Tsonkov N 2016 *Emerging regions and markets* (Sofia: UNWE Publishing house) p 251.
[11] Georgiev L 2014 *Regional Economics* (Sofia: New Bulgarian University) p 438
[12] Shishmanova M 2010 *Regional development and spatial planning. Research Theory and Methodology* (Blagoevgrad: Avanguard Prima Publishing house) p 1-23
[13] https://www.jptactis.com/Archives/jpt305/petrov.htm
[14] Devejiev M 1996 *Geography of transport* (Sofia: Forkom Publishing house) p 248
[15] Tsonkov N 2013 *Geoeconomic opportunities for increasing Balkan security and cooperation (*Sofia: International Relations Vol 1*)* p 37-44
[16] European Commission (2013), "Regulation (EU) No 1316/2013 of the European Parliament and of the Council of 11 December 2013 establishing the Connecting Europe Facility, amending Regulation (EU) No 913/2010 and repealing Regulations (EC) No 680/2007 and (EC) No 67/2010", Official Journal of the European Union, No 1316/2013, L. 348/129.
[17] Ezgeta D, Čaušević S, Mehanović M 2022 *Challenges of physical and digital integration TEN-T networks in Southeast European countries* (Transportation Research Procedia Vol 64) p 270-279
[18] Troch F, Sys C, Vanelslander T 2016 *Opportunities and challenges for rail freight: what can bet the future added value?* (Paper presented at the 4th Research Seminar on Railway Policy, Karlsruhe, 11 March 2016*)*
[19] Dunmore, Dick, Preti, Alberto, Routaboul, Clémence (2019) *The "Belt and Road Initiative": impact on TEN-T and the European transport system* (London: Journal of Shipping and Trade Vol. 4, Issue 10) p 1-17
[20] Elburz Z, Nijkamp P, Pels E 2017 *Public infrastructure and regional growth: Lessons from meta-analysis* (Journal of Transport Geography Vol 58) p 1-8
[21] De Fabiis F, Mancuso AC, Silvestri F, Coppola P 2023 *Spatial Economic impacts of the TEN-T Network Extension in the Adriatic and Ionian Region* (Sustainability Vol 15, Issue 6) p 1-15
[22] Stoychev K 2018 *The Regionalization of Bulgaria* (Public Policies No. 3 Vol. 9) p 1-22
[23] Dobrev D, *Bulgaria's regional policy in the context of de facto EU membership* (Varna: Science and Art, Economic University Publishing house) p146

IOP Conf. Series: Materials Science and Engineering 1297 (2023) 012019 doi:10.1088/1757-899X/1297/1/012019

On the need for connectivity and a new strategic approach to building road infrastructure in the Balkans

Nikolay Tsonkov[1], Kamen Petrov[2]

[1] Regional Development Department, University of National and World Economy, Sofia, 1517, Bulgaria

[2] Regional Development Department, University of National and World Economy, Sofia, 1000, Bulgaria

[3] kolio_tzonkov@abv.bg, petrovkamen@abv.bg

Abstract. The Bulgarian state is characterized by its geostrategic location. However, the country has not taken advantage of its natural localization. The main reasons for this are the lack of strategic thinking and implementation of a targeted and active transport policy within Bulgaria and the Balkans. This policy includes the planning and realization of projects in the field of strategic infrastructure. The policy is adopted in the context of the trans-European transport corridors. This will create conditions for connectivity between Bulgarian regions and conditions for stimulating cross-border cooperation. In this direction, the authors analyze the transport sector and infrastructure in Bulgaria to highlight the important areas where the Bulgarian state needs to take urgent action. In this regard, it is necessary to formulate the major priorities that will take place in a new strategy for the transport sector. The strategy is related to the reconducting of political activities, measures, and acceleration of the investments in Bulgaria's strategic road infrastructure.

1. Introduction

The post-Covid-19 world and the war between Ukraine and Russia started to acquire new contours and set new stereotypes and priorities. Thus, the Balkan countries began to become increasingly important zones and territories for transferring goods, people, and capital within the global economy. In addition, it turned out that in the new conditions, the logistics and the model of transporting goods and their delivery to the respective cities and settlements changed. For countries like Bulgaria, these problems have been exacerbated because our country has started to play a destructive role in slowing down the transportation of goods in the north-south and east-west directions. The events of long queues and the problems in moving freight within the Bulgarian national space once again highlight the need for a quality national transport strategy to guarantee our national interests, above all in economic terms. In addition, we must be aware that our country must integrate its transport system with that of neighboring countries such as Greece, Turkey, Romania, North Macedonia, and Serbia. The lack of national quality solutions to the transport problems that have arisen over the last 20 years have not found effective solutions and are now being repeated with greater poignancy as transport flows to increase and are increasingly important for business. What is worse is that our transport system remains in development, on the other hand, it also remains somewhat isolated from the European transport infrastructure and last but not least there is a lack of quality alternative routes on the main road routes in the country. The situation is similar in other countries, such as North Macedonia, Romania, Greece, and Turkey, but decisions are being taken quickly there, although there is a corresponding delay in the construction of road infrastructure. Thus, the objective of this paper can be deduced to outline some possible improvements in road infrastructure development and form a steep strategic plan to improve connectivity. Now is the time to address the issue of the region's transport links to the European transport system in the long term through a coordinated regional transport strategy.

This implies an analysis of the country's infrastructure situation and performance to outline possible strategic transport infrastructure projects. In terms of regional development, these projects become relevant not only in terms of connectivity but also in terms of the territorial and transport sector development in the EU. This approach means finding points of contact between the individual countries to achieve a high degree of cross-border cooperation and to develop a cross-border modern technological transport system of European importance for the countries.

The definition of regional priorities and infrastructure development projects should be based on a common interest. For the countries of south-eastern Europe, these are the corridors of national and regional importance that connect them to the transport network of the countries of Europe.

Currently, Bulgaria is precisely the strategic bridge that connects the war-affected countries of Southeast Europe with Western European countries and infrastructure. The decision taken at the European level for the construction of the Trans-European transport network (TEN-T) creates premises for the integration of different types of transport connections that will create better mobility between the individual parts of Europe. The philosophy of the network is to consist of two levels. The first is a "fundamental" or "core" network, with a clear construction horizon possibly until 2030, consisting of the strategic connections and transport hubs in the European Union. This deadline due to the coronavirus pandemic and the war in Ukraine may be extended until 2035 or 2038. The second direction "Broad" with a horizon of 2050, has the major purpose of ensuring availability and connectedness for European regions. The major European infrastructure network (TEN-T) passing through Bulgarian state space contains: – the routes of the Crete corridors – the Danube River inland waterway which is part of (the "Rhine and Danube rivers water route"); Vidin and Ruse ports at the Danube River (the corridor includes transport hub in Ruse); "Orient/Eastern-Mediterranean" corridor which contains: railway and road infrastructure from Vidin to Kulata that goes through the capital of Bulgaria; and the route of Sofia - Plovdiv – Burgas with second direction to Svilengrad which is located nearby Turkish border.

This new European policy gives reason to create new branches or renew the routes of the Crete corridors (Crete 1994 and Helsinki 1997). In practice, trust in the European strategic infrastructure roadmap affects the interests of many countries. In the period 2009-2019, the Bulgarian state is making attempts to improve its transport infrastructure as much as possible through the implementation of several infrastructure projects such as highways "Trakia", "Maritsa" and others, which will improve both the national transport system and transit through Bulgaria.

Over the years, strategies for the implementation of the "Three Seas" initiative and the Adriatic-Black Sea connectivity have also become urgent. Accompanied by certain difficulties in Bulgaria, the transport sector connections and road construction registered a decline in the period 2020-2023. All infrastructure facilities under construction, such as the highway section Sofia-Dragoman-Kalotina, as well as the commissioning of the Rudozem-Dimario border crossing, are slowly progressing. It turns out that neighboring countries Serbia, Romania, North Macedonia, and Greece have made significant progress in road construction at the expense of Bulgaria. This means that the main goal of our transport policy should be to construct a road infrastructure that forms the strategic bridge and connectivity of regions in Southeast Europe. In this direction, the search for an active transport policy on the part of Bulgaria with its neighbors is necessary, mainly in the reasoned different approaches to the realization of the European transport corridors and common infrastructure. In this direction, it can use the opportunity to build a corridor Reni-Izmail-Galac-Silistra-Turnu Magurele -Novi Sad-Dunayvaros. which will be important for the transfer of Ukrainian grain on the Danube River or a new economic impetus for the geo-economic development of cargo transfer on the Danube River. This is one alternative for good solutions to overcome the transport problems in the Balkan region. A similar important trans-regional project could be the road Galac - Braila - Ialomitsa - Culras - Danube-Silistra bridge, this will be a new expressway will connect 4 counties in Romania (with a length of 142 km), the Danube bridge around Silistra and in the future Silistra-Shumen expressway (111 km.). Similar projects will contribute to the development of transport policies and the solution of many issues related to regional development, transport systems, investments, and freight transfer.

2. Methodology of the research

The authors consider the transport sector as a key factor for regional and economic development. It is based on the concept that transport is an economic activity that is connected strongly to the infrastructure. Looking at that angle the transport delivers connectivity to regions. In this sense, the research methodology is dedicated to evaluating and analyzing European transport policies in the context of Bulgaria's integration into the European Union.

The major purpose of the paper is based on the transport sector and connectivity of the Balkan region analysis to highlight the contours of the future transport policy. In this term, the authors argue the necessity of a new strategic approach for planning road infrastructure in Balkan countries. Construction of the strategic infrastructure will provide a necessary level of connectivity to the countries themselves and among them (also cross-border connectivity). The authors refer to official statistical information that reflects ongoing spatial and descriptive processes and phenomena. The present article uses the network approach, which is necessary for an integral assessment of individual modes of transport and their framework in terms of their importance for the regional development of the Bulgarian state. The review of strategic documents and analyses, as well as the reference to an expert evaluation, allows us to determine the framework of the development of transport and the means that must be sought for the realization of transport networks. The geographical factor, as well as the economic one, must be considered to determine exactly where to invest the money to improve connectivity, quality of public transport, and quality of infrastructure. The offer of transport services by passenger cars can be evaluated by the level of motorization. Motorization is measured by the number of cars owned out of 1,000 people in the population. The trend is to increase motorization, both for the country as a whole and by regions. In some areas, motorization is ahead of the national average, such as in the Vidin district (703 cars per 1,000 people), Ruse region (606 cars per 1,000 people), Dobrich region (567 cars per 1000 population) and Sofia region (capital) – 525 (567 cars per 1000 population), and in other areas motorization is lower than the average for the country, such as Smolyan region (352 cars per 1000 population), Targovishte region (361 cars per 1000 people), Veliko Tarnovo region (369 cars per 1000 people population) and Pleven region (370 cars per 1000 people). The average motorization for the country is 441 cars per 1000 people For example, in response to a question about what is most important when using transport, some users answered that the level of comfort (23.0%), for others the level of awareness (18.6%), the degree of accessibility is a higher priority (15.4%), speed and security of transport services (25.4%), highway construction (14%), which shows that an important role in the analysis process is also the reference to sociological and economic research.

3. The transport sector in Bulgaria

Nationally and internationally, the transport industry is becoming increasingly important and involves more personnel, both directly and indirectly. The Bulgarian transport sector creates a high added value in GDP, and through its activities, it should ensure territorial and socio-economic cohesion of the country's regions by providing fast, convenient, and reliable transport. Its mission is to contribute to balanced regional development, integration of the most remote regions, and reduction of the number of overpopulated areas in urban agglomerations. Another essential task is to improve the quality of the transport infrastructure and maintain the high transport permeability of road vehicles in the national transport system. [1]. Public transport also contributes to ensuring that citizens have the same opportunities to reach jobs and cultural and entertainment venues. To ensure sustainability in its operations, the transport sector needs to improve itself through the introduction of innovative vehicle technologies as well as the improvement of adjacent infrastructure. This process requires investment in road development. However, in recent decades, analyses have shown a lack of investment activity related to infrastructure development. This obstacle plus increases in demand for transport services necessitates medium- and long-term planning. The long-term vision is set by the need for a modern and adequate road network, which requires a steady flow of financial investment. This process is facilitated by a high degree of ownership by all institutions and organizations involved in the sector. Government policy should be based on the principle of partnership, thus presumably stimulating the participation and role of private entrepreneurs in upgrading the communications network. This objective can be achieved by strengthening the functions of the private sector based on the realization of various forms of public-private partnerships [2]. The most relevant indicator for assessing the progress made in promoting the use of "greener" vehicles is the development of the indicator "Ozone precursor emissions from land transport". The total length of roads from the republican road network of Bulgaria as of October 18, 2022, is 19,907 km, of which 846 km are highways. The National Road Network (NRN) provides the connections between the settlements in the country, as well as those with the neighboring countries, for the transportation of passengers and goods. Roads in the national network are divided into different hierarchical classes according to criteria such as the economic and administrative functions they perform and their role in the overall transport system. In this sense, the following classes can be distinguished: trunk roads and first, second, and third-class roads. All of them create the conditions for connecting the different regions, thus forming a national road system. Some of the roads built are also of European importance because they are part of the trans-European infrastructure network. Thus, the roads in Bulgaria by 2023 have the following characteristics, first-class roads are 2,879 km, second-class 4,019 km, third-class 12,219 km. In this direction, we should note that 19615 km. are paved roads, 19,461 km are asphalt-covered, 32 km are paved, 122 km are ballasted, and 327 km are unpaved. Importance of infrastructure in driving economic recovery. To ensure competitiveness in an ever-globalizing market, it is important to address the need for new infrastructure development while preserving existing but aging road infrastructure assets, many of which need replacement, updating/upgrading, or repair. The total density of the entire road system is 0.39 km/km², which is lower than the EU average (0.51 km/km²), but higher than countries such as Poland, Slovakia, and Turkey and equal to Latvia, Lithuania, Romania, and Slovenia. However, the gap in the area of motorways is significant. The condition of many third-class roads is extremely poor. Because about 40% of Bulgaria's topography is mountainous and therefore with a lower population density, the overall density of the road system is sufficient. Most of the roads have not been repaired for more than 15 to 20 years, with scheduled periodic maintenance every 5 to 7 years and comprehensive maintenance and repair every 12 years. Most of the third-class roads mentioned above need reconstruction to meet the standards for such. Highway coverage throughout the country, with four or three-lane roads, is quite patchy. In Bulgaria, the main part of the road network is oriented in the east-west direction. It follows that the road connections in the North-South direction are poorly developed. The analysis of the authors shows that in the peripheral territories (especially along the southern border, the Danube territories, and those between the Cretan corridors 4, 7, and 9) the provision of transit services and the economic impact is insufficient [3].

Over the years, there has been a chronic shortage of financial resources for the renewal and rehabilitation of the road network. this is why the condition of roads is at a critical level in some regions. Usually, an important indicator that characterizes the quality of road infrastructure is the current condition of the road surface. It is analyzed and assessed by the developed and accepted methodology for establishing the degree of deterioration of the road surface [4]. This method is based on two criteria - type and extent of existing pavement damage. These indicators relate to the general condition of the road surface: or more specifically:
- good condition - surface damage of up to 10%;
- medium condition - 10 % to 30 % pavement surface damage;
- Poor condition - pavement surface damage greater than 30 %.

In recent years, a change has been noticed in intra-city passenger transport. In several smaller settlements, public transport was liquidated, but in large cities, over 40 thousand people remained, albeit on a smaller scale. It took practice to implement electric buses. Thus, their role becomes increasingly significant, because in the last quarter of 2022, a total of 72,516.0 thousand passengers were transported in urban transport, or 43.9% more compared to the same period of 2021. This is also the case with land transport of 30,804.7 thousand tons or 6.4% more compared to the last quarter of 2021. The work performed also increased by 22.4% and reached 12,527.1 million ton-kilometers.

Naturally, it turns out that with the modernization of public transport in certain areas such as Sofia-city, Plovdiv, Stara Zagora, Burgas, Shumen, and Haskovo, the demand for car freight services has also

IOP Conf. Series: Materials Science and Engineering 1297 (2023) 012019 doi:10.1088/1757-899X/1297/1/012019

increased. According to NSI data, the number of transported passengers by land transport in the fourth quarter of 2022 is 93,130.7 thousand, or 16.4% more compared to the same period of the previous year. The increase in bus transport is 14.8%, and in railway - by 44.0%. The work performed is 2,213.8 million passenger kilometers, which represents a growth of 40.3% compared to the fourth quarter of 2021. On the increased importance of transport. We can give an example of railway transport, which is in a difficult period. The main statistical indicators mainly include transported goods per ton - according to the type of transport - domestic (within the borders of the country) and international (imports, exports, and transit) is indicated in figure № 1, which shows that the railway transport is constantly increasing the transported goods.

	2021				2022			
	I	II	III	IV	I	II	III	IV
Transported cargo - thousand tons	4191.9	4642.2	4460.2	4775.0	4526.8	4886.3	4988.0	4890.0
including domestic transport	2688.2	2983.8	2824.8	3058.9	2908.5	3076.6	3213.0	3040.5
International shipments	1503.7	1658.4	1635.4	1716.1	1618.3	1809.7	1775.0	1849.5
Work performed - million km	1106.3	1188.2	1111.2	1251.9	1209.6	1284.3	1352.5	1392.6
including domestic transport	682.0	723.3	634.1	740.8	712.2	751.4	797.1	791.8
International shipments	424.3	464.9	477.1	511.1	497.4	532.9	555.4	600.8

Figure N1 Cargo transported and work performed by rail transport - Source: NSI

In practice, in Bulgaria, the transportation of goods not only by railways but also by road transport has seen significant growth and the transport industry has been brought to the fore. The demand for international freight transport services is determined by the country's foreign trade and transit flows, as well as by the competitiveness of Bulgarian transport companies. Unfortunately, the Bulgarian transport sector fails to be used effectively enough by EU policies for their financing. In this direction, most of the funding by the European Union (78%) for transport falls under common management so far. However, the share of funding managed directly by the Commission is increasing - from 13% for the period 2007-2013 to 31% for the period 2014-2020. The main reasons for the delay in the development of the transport sector can be found in the lack of sufficient ready-made projects, a quality partnership between businesses and the public sector, as well as the squandering of European funds on ineffective undertakings and projects. It is important to note that the car fleet in Bulgaria is constantly growing, although with a high average age of the cars, it finds it difficult to be competitive. The main road and railway network, as well as the ports for public transport with national importance and civil airports for public use, will remain and henceforth public state property. Some elements of the transport infrastructure, however, should be leased to the private operation sector to guarantee public services and non-discriminatory access to them. In a national aspect, road sections of regional significance will also be developed, which connect secondary and tertiary nodes to the TEN-T network. This means that Bulgaria should try to build a reliable intermodal infrastructure by 2030. The infrastructure required for intermodal transport consists of rail connections between major intermodal hubs. Railway links should be defined in such a way that they offer, compared to road transport, an efficient, safe, and competitive operation for carriers, and the selected network nodes should allow intermodal companies to handle large volumes of freight and divert freight for rail transport services. This means the implementation of several significant cross-border projects and the implementation of joint initiatives between the Bulgarian countries. For example, work can also be done to improve rail connectivity.

For example, to modernize the following railway connections we need strategic investments because of the necessary resources. These are lines in Bulgaria, included in the European Agreement as important lines for international combined transport and related facilities (AGTC). These lines are Bucharest - Ruse - Gorna Oryahovitsa - Dabovo - Dimitrovgrad - Avren - Komotini (450 km), and from Dimitrovgrad to Komotini must be built. - Sofia - Mezdra - Gorna Oryahovitsa - Kaspichan - Varna (543 km), the line must be completely renovated Nis-Tsaribrod- Dragoman - Sofia - Plovdiv - Dimitrovgrad - Svilengrad (482 km) - Plovdiv - Zimnitsa - Karnobat - Burgas (294 km) - Vidin - Sofia (279 km) - Sofia - Kulata - Seres-Thessaloniki (350 km). The interest may also be the attempts to build reliable intermodal nodes under the European Agreement [5]. In this agreement or the map of the extended network, except for the directions and nodes in the existing TEN-T map, the following new proposals are also included: five new road routes – direction Sofia - Veliko Tarnovo - Shumen - Varna, Rila Expressway, along the direction Kyustendil - Dupnitsa - Samokov - PV Bogoroditsa - Highway "Trakia" / Highway "Hemus", Expressway I-2 "Ruse - Shumen", Expressway "Varna - Durankulak" and Road "Plovdiv - Rudozem", Port Silistra. It also includes the railway connections to Varna. The construction of new border crossing points of the Republic of Bulgaria with North Macedonia, Greece, Turkey, Serbia, and Romania is also important for transport. It is also necessary to invest funds for the total modernization and robotization of the Dragoman, Vidin, Kulata, and Ruse checkpoints, which will have a transit character if Bulgaria joins Schengen. Such an important border remains the one with Turkey, which needs at least two more checkpoints, as well as a total modernization of the Svilengrad border crossing. It is necessary to move the Container Terminal in Sofia, to Pernik station by doubling the Sofia – Voluyak - Pernik line. Another solution is to build a similar terminal around Elin Pelin station. With similar infrastructural solutions, the countries will improve their capacity and opportunities for the development of the regional economy. Big projects in the field of transport require massive investments and their implementation covers long periods. Once operational, it is likely that these projects to have a significant impact on the performance of the network of transport for decades, both at the EU and Member State level. Therefore, such projects must be planned as part of a strategy or a broader plan regarding transport infrastructure.

4. The need for an active policy to improve connectivity and transport development

In terms of transport, the active policy of the Bulgarian state should be aimed at initiating and building joint border and infrastructure projects with North Macedonia, Turkey, Serbia, Romania, and Greece. In the next stage, work should be done on the European transport system with countries such as Croatia, Bosnia, Albania, and Hungary. In this way, the design and construction of regional strategic transport links can have greater support. The mistake of the Bulgarian state is the lack of priority for the preparation of infrastructure projects with several countries, for example, the creation of a water connection between Silistra - Turnu Magurele – Belgrade - Novi Sad - Vukovar. We can outline other projects as well. Such initiatives will lead to increased attention to South-Eastern Europe and will help the region attract new investors due to improved transport accessibility [6]. Here it is necessary to emphasize that the regional strategic transport project must be based on modern technological solutions for all elements of the transport system and the transport process, namely:

- the solution to the key issue - that the third Danube bridge between Bulgaria and Romania becomes a joint initiative between Hungary, Bulgaria, Greece, and Turkey. [7]

- Creation of a joint administration between Bulgaria, North Macedonia, Romania, and Albania for the construction of Corridor VIII and its extension to the city of Constanta in Romania. - The construction of the transport infrastructure along Corridor VIII and the Bourgas - Batumi ferry in the Black Sea.

- The harmonization of legislation and technology for border crossing by European standards will ensure the efficient functioning of the transport system.

- construction of reserve airports of national importance near the cities of Vlora, Kumanovo, Plovdiv, and Tulcha.

For the general development of the Balkan countries, Corridor No. 4 is of essential importance, which plays a role in connectivity between the Western Balkans and connectivity to Turkey, and the Middle East. Pan-European Transport Corridor IV is one of the Pan-European Transport Corridors. It starts from Dresden and Nuremberg in Germany and ends in three cities: Thessaloniki in Greece, Constanta in Romania, and Istanbul in Turkey. The corridor to the west starts from two points - Dresden and Nuremberg. The branches connect in Prague and go in the direction of Vienna - Bratislava - Gyor - Budapest - Arad. From Arad, the road branches off in two directions. One direction continues to Bucharest and Constanta. The second direction is in the direction of Craiova - Sofia. From Sofia, the southern branch leads to Thessaloniki, and the eastern branch to Plovdiv and Istanbul. The corridor is the shortest land connection between Greece and Central Europe and is located entirely within the EU. There is hardly any need to waste words on how important transport links are for the functioning of the European Union. The situation in this area affects almost every citizen's life and is directly related to the four fundamental freedoms (movement of people, goods, money, and services) that are the basis of the EU. That's why connecting Europe's roads into one whole was a priority issue that was worked on intensively even before the EU adopted its current form, and revisions of plans and rules are constantly being made. In a national aspect, road sections of regional significance will also be developed, which connect secondary and tertiary nodes to the TEN-T network. The quality of the transport service is evaluated using a utility function. The utility for rail freight customers is presented as a function of the following indicators: speed; periodicity; value; capacity; flexibility; availability; reliability; and rity. Another important point on which the capacity of the road sections depends is the class of the road, road conditions, regulatory speed limits, and technical requirements of the motor vehicles that move on the elements of the road network. In this direction, in Bulgaria, it is accepted that the quality of the road surface is low and does not meet good standards. Another issue is that there are not enough reliable tools to assess the quality of roads, and that is where the significant problems come from. The port facilities in Varna built at the beginning of the last century also proved to be a significant problem for transport (East terminal of Varna port), Burgas (East terminal of Burgas port). The location of the ports in Ruse, Svishtov, Lom, Oryahovo, and Tutrakan is in the central area of the cities, which is a significant obstacle both of an ecological, and architectural-construction nature. Also, there is a need for their rehabilitation and modernization. Finding the right formula and balance of state and municipal interests with the public need for a new and ecological urban environment requires rethinking the concept of the forms of use of these territories.

5. Urgent measures on international corridors

It is important to note that Bulgaria is very slowly and partially developing its infrastructure, especially in terms of transport corridors. Moreover, after 2014, the corridors remained on the second line in comparison with the modeling of the European transport system. This will most certainly happen to the Bulgarian state administration after 2025 when it will have to hurry with the infrastructure and the management of European funds. A similar development is with the construction of the route of corridor No. 8 from Sofia-Voluyak-Pernik-Radomir-Kyustendil. In addition, the fate of the doubling and electrification of the Dupnitsa - Kulata railway line is not clear. The problems with highway "Struma", the modernization of the railway line Plovdiv - Svilengrad, the completion of the motorway to Turkey, the doubling of the railway line from Chirpan to Sliven, the improvement of transport connections with Greece, and other infrastructure objects that have not yet been built. A third, and why not a fourth, bridge between Romania and Bulgaria is important and necessary. Especially for the bridges on the Danube River, they should be joint projects of at least Romania and Bulgaria, not to mention including countries such as Moldova, Greece, Turkey, and Ukraine. It should be clear that we could hardly fight for such projects alone without the participation of other interested parties. Moreover, our northern neighbor would hardly allow the transit transport market, which has grown significantly in the last 10 years, to be taken away from her. Over the last 6-7 years, while we have been talking about transport corridors mainly at responsible international meetings, Romania has developed its transport infrastructure in the Black Sea area to a very large extent by putting into operation ferry services to

Georgia and Turkey, increasing the capacity of the port of Constanta, introducing modern technologies for transporting goods, developing combined transport by rail to the countries of Central and Western Europe. In short, the transport infrastructure along the Corridor 4 wing through Romania was prepared before us for increasing the transit transport flow based on infrastructure projects and technologies. In this situation, there is progress in the infrastructure of Serbia, Montenegro, North Macedonia, Albania, Turkey, and Greece, and our infrastructure development is not proceeding at the required pace. Some projects, for which funding has been secured under European programs and the World Bank, are delayed for years due to both the borrowers' fault and the bureaucratic machine. Much better management coordination is needed to compensate to some extent for the lag. It is necessary for our country to fully realize at least one European corridor within our country. At this stage, it seems most realistic to me to implement Corridor No. 8. It is a connecting axis from the Adriatic to the Black Sea and a gateway to the region of the Caucasus and Eastern Asia. This project, together with the construction of road infrastructure along corridors IV and IX, will give Bulgaria an opportunity for regional leadership and modern connectivity in the directions north-south and east-west. The transport strategy of the country under the current circumstances should not be limited to the European transport system, it should set a new horizon for the entire Bulgarian territory. This means that infrastructure projects and technological interaction between the various entities of the government and business should be implemented as a priority. Thus, for Bulgaria and its neighbors, it is necessary to technologically renew transport and create opportunities for integrated management of transport between all countries of the Balkans. Bulgaria and the region of Europe and Asia need a transport technology system with modern infrastructure and advanced transport technology. In this context, transport infrastructure in general, and transport technology should be seen as a strategic bridge for the integration of Bulgaria and the countries concerned in the region into the European transport system. Also, through the quality of the information environment, the resolution of the problems facing the Bulgarian port system will be achieved. In one way or another, this will have a positive impact on safety and security. Investing in the expansion, modernization, and renewal of port infrastructure, as well as in new technologies and equipment, will increase the reliability, availability, and security of the handling of vehicles and the unloading of goods. Several ports, no new quay machinery has been purchased for many years [8]. The main goals and tasks to be performed in developing the future design solutions for the development of transport schemes and technologies in port terminals should be directed towards technological equipment being at a modern technical level, yes meets the conditions and volume of work. Maintaining the design depths in the ports is of particular importance for commercial transportation, but it is also the basis for transport safety - avoiding accidents accident as "bottoming out of the ship" or the even more severe "grounding". Incidents such as these can have major consequences for the ship itself, the port, and the country concerned. The risk of damage to the integrity of the ship also poses environmental risks. Such emergencies can have a strong negative impact on the reputation of Bulgarian ports. It is therefore important that they are perceived by maritime transport actors as a safe and secure area [9]. All these problems, which are characteristic of the development of the Bulgarian transport corridors, lead to the increase of problems in the movement of goods, people, and capital. This is particularly evident from the dramatic increase in transport volumes, which may lead to capacity challenges for some modes of transport in the EU. Congestion is already a serious environmental and economic problem, currently costing the EU nearly €140 billion every year.

6. Construction of cargo terminals and logistics areas - concept and possibilities

The European Strategic Framework for Transport comprises a series of documents whose main objective is to build a common transport area, a high degree of transport connectedness, smart and sustainable mobility, a significant reduction of harmful emissions, an increase in the competitiveness of the transport sector to achieve regional economic development. The major documents are - the Transport Strategy of the European Commission [10], the Transport Lending Policy 2022 of the European Investment Bank [11], and the Transport Sector Strategy (2019 - 2024) of the European Bank of Reconstruction and Development [12]. An important milestone in the development of the European transport sector was the publication of the White Paper on Transport. These recent documents build on the Paper, shaping EU policy in the long term. The basic concept is to build transport networks (roads, railways, airports, sea, and river traffic) as well as energy networks, GA, and electricity. These networks must be connected to logistics centers and settlements that provide the necessary stock availability of the goods for the major markets [13]. An important document that also identifies transport sector weaknesses and challenges is the European Court of Auditors report [14]. It analyses the current state of transport, identifies the problems, and proposes solutions to overcome the constraints facing the sector.

European strategic documents determine the policy of each nation-state (including Bulgaria). In a hierarchical order for the full integration of the Bulgarian transport sector in the common space, some key documents have been adopted, such as an integrated strategy in the period up to 2030 [15], a National plan for the development of the combined transport of the Republic of Bulgaria up to 2030 [16], and Programme for the development and operation of the railway infrastructure 2020 - 2024 [17]. In the current year 2023, the Programme for the Development of the Railway Infrastructure and the Programme for the Development of the Railway Infrastructure will be adopted. The main problem that exists in Bulgaria is related to the fact that all adopted strategic documents are not realistic enough, are not implemented or only partially implemented, and there is no financial mechanism for their implementation. The construction of similar areas and centers will ease the burden on cities and will make business processes known in more peripheral areas. In this direction, the location should be optimally close to the railway network, the highway network, or the waterway network, away from populated areas to cause as little inconvenience as possible. Built on similar hubs and zones can improve the quality of the transport network and meet the needs of users (considering peak hours of car traffic, railway timetables, port dwell time restrictions, etc.). The related services must be:
- Trans-European transports.
- Offices for operators.
- Warehouses or depots for ITU (empty, loaded, dangerous, frozen, etc.).
- Parking for road chassis.
- Maintenance facilities (repairs, laundry, etc.).
- Proximity to a logistics center (warehouses, etc.).

The capacity and quality of the transport network should be appropriate to the needs of users (co-side ring peak hours in road traffic, rail timetables, port dwell time restrictions, etc.).

The European Union's transport policy is also aimed at counteracting the demographic pressure on the territory, including the release of pollutants into the atmosphere. In this context, the policy promotes the use of greener modes of transport. There is therefore a growing view that EU countries need to implement targeted measures to invest in urban and rail transport, to develop river and sea routes. The key objective is to promote short sea shipping linked to transport hubs enabling transport connectivity. Clean transport zones are considered one of the most important tools in the fight against toxic emissions from road transport. Research by the Clean Cities Campaign shows that zoning can also have a positive impact on local businesses. Opponents of these business zones say the zones will kill businesses by reducing the number of drive-thru customers. However, analysis of several European cities' Clean Cities campaigns from 2021 showed that the zone does hurt business, quite the opposite. Research from Madrid showed that pre-Christmas spending in the SCT city center grew faster than in areas outside the zone - 8.6% versus 3.3% respectively. Similar studies in Bucharest and Athens showed no effect of zones on the number of people visiting the shops. In addition, a review of studies confirmed that people who bike or walk to stores tend to shop more often [18]. Only a few changes are indicated render convention are structures unable to meet the growing needs of cities, and existing infrastructure has reached the limits of its capacity. Overcoming logistical problems therefore requires the search for new concepts and business models that enable sustainability and low-emission urban development. Therefore, it is necessary to identify new logistics solutions that will not only enable the efficiency of cities but also the restoration" of the space in the city, previously used by vehicular traffic. The trends observed so far indicate a state of imbalance and growth load on the natural environment caused by logistics, the subject of which is the spatial configuration of transport and storage networks that allow continuous following of the city as an economic and social space and cultural. The modern, globalized world increases the importance of logistics constantly is growing. The negative effects of its impact on the surroundings and the environment are also increasing [19]. Thus, in the new conditions, transport cooperation in the Balkans may have its characteristics modeling the economy and development of this region. It is important to note that the so-called zones and logistics centers can play smaller centers that have an important focus such as Kumanovo, Ruse, Silistra, Edirne, Gevgelija, and others. The goal is f, or these cities to be important and strategic points that can provide resources to the big cities and areas near them. This will allow European transport corridors to have another significant place in the European transport system of centers for socio-economic development and core [20].

The development of corridors is a matter of solidarity. This commonality requires the synchronization of the transport systems of the individual European countries, including Bulgaria. Therefore, the objective need for a high degree of convergence in the European network to achieve a high level of sustainability and strengthen the single market comes to the fore. The transportation system must be efficient in terms of transport operations and their users. In particular, the advantages within the transport corridors should be compatible with the system; the delivery must exceed costs. These objective measures the benefit provided to users and service providers in the transport system and the quantitative measures of it are cost-benefit ratio (RBC), net present value (NPV), and internal rate of economic profitability (RIRE) [21]. In this direction, our analysis will go beyond the management model of these logistics zones, because in practice they must perform economic and other logistics-transport functions. This means that within the framework of the transport corridors, the cities, and settlements through which they pass should not be neglected, but those spatial solutions should be sought for the construction of peripheral urban areas and logistics centers that are close to the main highways and routes. but not to complicate life in the cities and above all to respect the relevant ecological needs. Building the transport system and modeling common transport policies through the transport corridors is the right policy for us. In practice, however, building this connectivity is a prerequisite for cities and their future. Realization requires a change in the way of working and the fulfillment of logistical goals. Conventional logistics structures for reconstruction and replacement. They should not be based on the exhibition or the intensive infrastructure. A new transport policy and spatial approach must be used. As an opportunity for a common concept based on the sharing or cooperation of interests. In our country, there are intermodal terms, such as Intermodal Terminal Plovdiv (IMT Plovdiv) is in the South-Central region of planning (YUCRP) in Bulgaria, Plovdiv district, "Rodopi" municipality, land of the village of Zlatitrap, Kamishta area, north of Todor Kableshkov station. The Todor Kableshkov train station is located at the first main railway line Sthe Sofia - Plovdiv - Svilengrad. The distance from the terminal to the city agglomeration is 4 km. The city of Plovdiv has turned into a transport hub in the TEN-T network. Intermodal terminal "METALIMPEX" Ltd. - Stara Zagora, has 5 tracks (5, 6, 7, 8, and 9 tracks from the freight taxation) with a length of 360 m., The total area of the terminal is 240 acres of concrete a site that is fully illuminated. The terminal has two gantry cranes, 2 unit locomotives, 3 forklifts, and 20 trucks with a sense of carrying out transportation in the country. The terminal can handle about 16,000 containers per year. European project funding is planned to increase the cargo turnover. It includes the construction of an intermodal terminal in Ruse, which will stimulate the economic and transport development of the entire planning area. This project must be seen in the context of the overall completion of the European transport network. The aim is thus to improve connectivity between the different modes of transport in the south-eastern EU region. At the same time, the Ruse project has created the conditions for better quality transport services. Work is also underway to build an intermodal terminal around the city of Varna. The aim is to improve intramodality in this region of the EU, by creating conditions for optimal interaction and integration of the various types of transport. As a result of the implementation of the project, the provision of the possibility of efficient transportation of large volumes of cargo from the Black Sea to Central Europe and trade hubs in the EU. The intermodal terminal in Burgas is owned by "Despred" AD. The multifunctional one-warehouse tone-warehouses 5 covered warehouses for general cargo storage with a total area of 5,980 sq. m and 12,00 sq. m open species.md storage areas. Two of the warehouses have suspended local load-lifting equipment (electric hoists, item 3t.) and have permission for temporary and customs storage of goods [15].

Conclusion

The war in Ukraine has caused serious shocks to the economic development of the European Union. At the same time, it has created opportunities from which European countries, particularly in Eastern Europe, can benefit. In this context, many North-South and East-West transport flows have been redirected through Bulgaria. However, Bulgaria has proved to be unprepared because its infrastructure is in a poor state. In addition, the Bulgarian state slept for many years and has not yet completed any of the five trans-European transport corridors. These problems severely limit the development of the transport sector in Bulgaria and, more broadly, the economic development of the country. The transport system includes in its scope all types of transport for general and special use, which can be combined in the process of operation and organization, to achieve economic, social, and environmental effects. The alternative nature of a combinatorial transport solution embedded within the limits of the transport system forms a potential for the development of the sector in unison with the pillars of sustainability, which aims to preserve the environment by reducing harmful human impact on living and non-living nature. The Bulgarian state must therefore take urgent measures to update transport policy and invest in infrastructure development, including in border regions. These measures are related strongly to the TEN-T network. This means that the Bulgarian government must use a strategic approach in planning road and railway infrastructure. The new approach is based on principles such as intramodality, connectivity, integrity, and networking. Additionally, the strategic infrastructure must include common projects with North Macedonia, Greece, Romania, and Serbia. This will create conditions for the development of the Bulgarian regions and improve connectivity between the Balkan countries.

For all of this, Bulgaria must pursue an active and targeted transport and foreign policy that will support the Bulgarian economy, especially the transport sector. The transport policy can be separated into two directions. The first is related to strategic infrastructure construction. And the second one contains measures for stimulating the transport sector and logistics. In this regard, it is important to build freight settlements and to accelerate the policy of construction and rehabilitation of road infrastructure. Important in this respect is the design and construction of more bridges on the Danube, tunnels under the Balkan Mountains (at Petrohan and Beklemeto), logistic intramodal hubs, and port equipment. In this term, the Bulgarian state must extend the capacity of the Black Sea ports (Varna and Burgas). These measures should be implemented in parallel with the modernization and construction of new railway routes, such as Bulgaria - North Macedonia.

Acknowledgment

The paper is financed by a scientific project of UNWE number NI 5/ 2021.

References

[1] Georgiev L 2014 *Regional Economics* (Sofia: New Bulgarian University) p 438

[2] Hristov H, Daskalova T 2011 *Infrastructure project management* (Sofia: Stopanstvo Publishing house) p 180

[3] Koster R, Petrov K 2020 *Infrastructure Corridor 9 as a factor for improving transport accessibility in Bulgaria* (*Geopolitics* Vol. 2) p 74 – 85

[4] Grozdanov H. 2019 *Methods for estimating external costs caused by road traffic accidents (Proc. of XIIth National Conference on Transport Infrastructure with International Participation)* p 44-56

[5] European Agreement on Important International Combined Transport Lines and Related Installations (AGTC). Geneva, 1 February 1991

[6] Dobrev D, *Bulgaria's regional policy in the context of de facto EU membership* (Varna: Science and Art, Economic University Publishing house) p146

[7] Grozdanov H 2021 *Need for connectivity between Bulgaria and Romania across the Danube (Proc. of XIVth National Conference on Transport Infrastructure with International*

Participation) p 111-114.
[8] COM(2001)264 – Communication Sustainable Europe for a Better World: A EU Strategy for Sustainable Development (Commission's proposal to the Gothenburg European Council)
[9] Stoychev K 2018 *The Regionalization of Bulgaria* (Public Policies No. 3 Vol. 9) p 1-22
[10] European Commission 2020 Sustainable and Smart Mobility Strategy – putting European transport on track for the future.
[11] European Investment Bank 2022 Transport Lending Policy: The Way Forward.
[12] European Bank of Reconstruction and Development 2018 Transport Sector Strategy (2019 - 2024).
[13] COM(2001)370 - White paper - European transport policy for 2010: time to decide
[14] European Court of Auditors. (2018). Building a successful EU transport sector - what are the challenges?
[15] Ministry of Transport and Communications. (2017). Integrated transport strategy.
[16] Council of Ministers. (2023). National Development Plan for Combined Transport in Bulgaria until 2030.
[17] Ministry of Transport and Communications. (2019). Railway Infrastructure Development and Operation Programme 2020 – 2024.
[18] Parks, K., Denholm, P. & Markel, T. (2007). Costs and Emissions Associated with PlugIn Hybrid Electric Vehicle Charging in the Xcel Energy Colorado Service Territory. Springfield: MMidwestResearch Institute.
[19] Rodrigue, J. P., Comtois, C. & Slack, B. (2006). The Geography of Transport Systems. London & New York: Routledge
[20] Szołtysek J., 2009, Podstawy logistyki miejskiej, Wydawnictwo Akademii Ekonomicznej w Katowicach, Katowice.
[21] Zołtysek J., Brdulak H., Kauf S., 2016, Miasto dla pieszych. Idea czy rzeczywistość, Texter, Warszawa.

Application of artificial intelligence in road network inventory and network-wide road safety assessment

I Katsarov[1], S Penkov[2]

[1] Federation of Highway Engineers, Sofia, Bulgaria
[2] Bulgarian AI Cluster, Sofia, Bulgaria
iv.katsarov@gmail.com, sv.penkov@gmail.com

Abstract. This research paper examines the application of artificial intelligence (AI) in road network inventory and network-wide road safety assessment. An overview of current research related to this field is presented and various aspects of the use of AI in road infrastructure and safety assessment are analysed. In today's world, where the road network continues to develop and grow, the use of AI can offer significant advantages in the inventory and assessment of road infrastructure and safety. AI can be used to process large volumes of data collected from various sources such as geographic data, photos, documentation and more. One of the main aspects that the article looks at is the use of AI to automatically process data and create a digital inventory of the road network. AI systems can analyse geographic data and photos to identify and classify roads, road signs, markings, and other elements of road infrastructure. This can facilitate engineers and competent authorities in the process of road inventory and maintenance. Another important aspect discussed in the paper is the use of AI for the general assessment of road safety. AI models can analyse road infrastructure inventory data, as well as traffic data, crashes, and other factors, to provide a comprehensive assessment of road network safety. This allows the competent authorities to identify vulnerable places, carry out preventive measures and plan appropriate interventions aimed at improving road safety. In conclusion, the paper highlights the possibilities and advantages of using artificial intelligence in road network inventory and network-wide road safety assessment. Such an approach can contribute to more efficient management of road infrastructure and improvement of road safety.

1. Introduction

In recent years the application of AI in almost every domain of world economics is already a fact. Transportation and road infrastructure management are no exclusion. The fact that the sector is one of the less digitalized, regarding the overall level of the digitalization of all other domains of the economy [1], drives the necessity of forcing the new digital processes to keep up with the rapid development and emerging needs of every stakeholder. There are already autonomous cars, that are driving around the road network, and we should make everything to keep the road assets on the highest level to keep the driving environment in the best condition possible, so we can guarantee safety and comfort of every user. Therefore, new technologies and AI can be used to force the speed and quality of road network inventory [2] and make it with higher pace and less subjectivity.

One significant area where AI is making a substantial impact is in road network inventory which can be of great help to the network-based road safety assessment. The integration of AI technologies is on the way to revolutionize this field, offering more efficient, accurate, and data-driven results.

On the other hand, road safety, particularly in Bulgaria experiences a variety of deficiencies. They are discussed in this [3] study. The external losses from fatalities result in billions of Euros and the exact numbers are calculated in a recent study[4],[5]. Therefore, it is so important that new technologies like AI and Deep Learning are introduced and applied to leverage road safety to a new higher level.

2. Methods for road asset inventory

Traditional methods for road asset data collection can be separated into 3 main groups – manual, semi-automated and automated. Each of them has its differences and they are usually based on factors like reliability of the data collection process and/or equipment. Different types of equipment is characterized with different level of accuracy and efficiency. The accuracy of the data obtained is crucial for the decision-making process after. And the efficiency of the data collection reflects over the quantity of data for the road assets that can be collected to facilitate the management system. Another important aspect is the cost for the data collection. Every method has its own specific, which include people, time, and equipment, which form the overall cost for data collection.

2.1. Manual collection

Manual inspection for data collection is the most traditional method for road asset data acquisition. It is detailed and accurate. It relies on experts' labor and requires GPS stations or total stations to gather information for location of assets. It is suitable for small-scale data collection, where other methods are costly to implement. It allows experts' decision making and makes the process more sophisticated. It is suitable for limited technological infrastructure.

On the other hand, this method is time-consuming and requires dedicated and well-trained experts. It is subject to human factor and therefore potential errors can occur. It is not applicable when there is limited time to collect wide amounts of data [6]. After data-collection the processing of data is something that makes the process even more time-consuming.

2.2. Mobile Mapping System (MMS)

The concept for mobile mapping system dates back to 1970s, where it shows a system, which integrates sensors and other equipment , mounted to a vehicle, in order to facilitate the task to collect georeferenced photos, that can be later manually analysed by trained experts, in order road assets to be positioned by their coordinates [7].

More recent technologies introduce a variety of other equipment like cameras, laser scanners, gyros, GPS, computers, storage devices, etc. They have high data accuracy, which is less laborious. This makes the process possible for more frequent applications. Data can be processed in the comfort of the office and requires minimum fieldwork. It is direct and accurate in every axis and gives very detailed information for positioning, distances, and sizes of the recorded asset data. Moreover, it is characterized with minimum hindrance to road-users [8].

The other side of this technology is that the initial set-up and equipment costs can be significantly high, because of the expensive equipment that is needed to facilitate only one vehicle. This makes it more complex and costly to be applied on a large scale. Another disadvantage is the big data that must be handled, stored, and analyzed [9]. On large-scale road network it makes it complex to apply and therefore becomes more suitable for separate projects and road design, rather than entire road network inventory. Most of them apply semi-automatic models for inventory.

2.3. Aerial surveys

Aerial surveys are wide-spread technique for mass data collection. In the recent years they became standard tools for road designers, construction companies, authorities, and others. They can be used to collect variety of data, based on the type of sensors the aircrafts are equipped with, they can collect topographic, geometric and imagery data. Aerial surveys can be conducted using a variety of aircraft, including airplanes, helicopters, and drones. The type of aircraft used will depend on the specific needs of the project.

The application of aerial surveys in road asset inventory offers several advantages. One of the main advantages of aerial surveys for road asset inventory is their efficiency. Aerial surveys can cover large areas in a relatively short amount of time, allowing for the inventory of road assets over a wide geographic area. Another advantage is the ability to capture detailed and comprehensive data. Aerial surveys can provide high-resolution images or data that capture the condition of road assets in detail. This level of detail can help road agencies assess the condition of assets such as pavement, signs, and markings, and prioritize maintenance or repair activities accordingly. Additionally, aerial surveys can also aid in the identification of new or hidden road assets that may not be easily visible from the ground. Furthermore, aerial surveys can enhance safety during the inventory process. By conducting surveys from the air, road agencies can avoid the need for personnel to be on the ground, which reduces the risk of accidents or injuries [10].

However, aerial surveys also have some limitations. One limitation of aerial surveys is the cost associated with conducting them. Aerial surveys require specialized equipment, such as airplanes or drones, as well as skilled personnel to operate the equipment and analyze the data. Another limitation is the dependency on weather conditions. Aerial surveys are subject to weather conditions, such as strong winds, rain, or fog, which can limit or disrupt the survey operations. Additionally, aerial surveys may face challenges in accurately capturing road assets due to occlusion caused by trees and buildings. This can make it difficult to extract road networks or identify assets that are obscured from the aerial perspective. By utilizing aerial surveys in combination with ground-based surveys, road agencies can benefit from the efficiency and comprehensive data capture of aerial surveys while also addressing the limitations of occlusion and cost by using ground-based surveys. Most of them apply manual models for road asset data collection and therefore are time-consuming.

2.4. Remote Sensing and Satellite Imagery

Remote sensing and satellite imagery have become more and more often regarded as instruments for ground surveying. They are performed by capturing high-resolution images from satellites and use of other sensors to obtain information for ground and objects. There are a lot of emerging providers that offer such service [13]. The technology has its benefits and disadvantages. This method offers wide area coverage and is suitable for large-scale and national road networks inspection. It can provide a historical database for before-after analyses and there is no need for physical presence. It is a high-performance method. On the other hand, the data is characterized with low resolution, and this makes it less accurate and not so reliable, when coming to decision making. Most of them apply manual models for road asset data collection and therefore are time-consuming.

2.5. Mobile Apps and Smartphones

With emerging technologies, smartphones became powerful instruments for everyday tasks and not only that, but also introduced them into areas like road asset management and data collection. Their advantage is that they are equipped with cameras, GPS, gyro, and accelerometers, etc. They are easy to use for field personnel and cost-effective for wide-scale applications. They have limited accuracy for certain attributes, compared to specialized sensors.

2.6. Connected Vehicles

Connected vehicles are often regarded as the future of transportation, but not only that. The variety of high-quality and resolution sensors, that are equipped with LIDAR cameras, GPS, and numerous other sensors. They can collect and process terabytes of information, which can be of use and analysed by road authorities to monitor road assets and road conditions. However, this technology is yet to develop.

3. Application of artificial intelligence and automation of road network inventory

3.1. Evolution of AI

The evolution of AI and automated detection for road asset inventory has been rapid in recent years. Early methods for road asset inventory relied on manual data collection, which was time-consuming and error prone. However, the development of AI and automated detection techniques has made it possible to collect road asset data more efficiently and accurately.

One of the earliest applications of AI for road asset inventory was the use of image recognition to identify road defects. This technique was first used in the early 2000s, and it has since been refined to the point where it can now be used to identify a wide variety of road defects, including cracks, potholes, and missing lane markings. [11]

In recent years, there has been a growing interest in using AI for more complex tasks related to road asset inventory, such as pavement condition assessment and traffic flow analysis. These tasks require a deeper understanding of road conditions and traffic patterns, and they can be more challenging to automate. However, AI techniques are being developed that can be used to perform these tasks with increasing accuracy.

The use of AI for road asset inventory is still in its early stages, but it has the potential to revolutionize the way road assets are managed. By automating data collection and analysis, AI can help to make road asset inventory more efficient and accurate. This can lead to better-maintained roads and reduced road accidents.

3.2. Application

A recent proposed framework [11] improves the repeatability and cost-effectiveness of the asset inventory process by integrating advanced artificial intelligence technologies for automated real-time detection, classification, and localization of traffic signs from roadway images. This reduces the need for subjective and time-consuming human interventions.

The model uses an edge device with the trained Machine Learning algorithm and the accompanying code, referred to as a plug-and-play device (Google Coral USB Accelerator) that runs a speedy deep neural network on each collected image. This effectively minimizes time between data collection and data delivery.

Moreover, the quality control process can be used to further train the machine learning algorithm and improve its effectiveness and reliability as more data is collected. This reduces the need for re-work and cost overruns, thereby enhancing the cost-effectiveness of the inventory process. Thus, the framework is designed for real-time processing and superior effectiveness, efficiency, and reliability.

3.3. Enhancing the data-driven decision-making process

The real-time processing capability of the proposed solution greatly enhances the data-driven decision-making process by accelerating the delivery of processed data. Conventionally, there's significant time lag between data collection and data delivery, but the proposed solution reduces this gap through real-time automated data processing.

This timeliness not only improves efficiency but also ensures that decision-making is based on the most current data, which is critical for accurate and relevant decisions. Furthermore, it reduces the data collection costs on large infrastructure assets network, resulting in cost savings.

In addition, the solution's capability to identify and classify assets (in this case traffic signs) in real-time minimizes the need for human interventions, which are often time-consuming and costly. This contributes to higher repeatability and cost-effectiveness, further aiding in more efficient, effective decision-making.

3.4. Deep learning

Neural networks were among the first machine learning models proposed in the late 50's, but recent advances in computing hardware have resulted in a Cambrian explosion of state-of-the-art models for computer vision and natural language processing. In particular, deep learning [12] focuses on building model architectures with millions or billions of parameters structured in a deep hierarchy of layers. Convolutional neural networks are one of the most prevalent model types within the computer vision domain, where convolutional layers learn visual features similar to the human visual cortex.

A key step of building a successful solution based on AI is the acquisition and curation of a dataset representing the task that's to be automated. Typically, this is the most resource intensive step where raw data samples (images in the case of computer vision) need to be manually labeled. Depending on the task this would mean manually indicating the bounding box around objects of interest (e.g. traffic signs) or drawing pixel masks (e.g. over lane markings). Importantly, gathering this dataset enables the complete automation of the task at hand and so novel data can be processed without the need for human intervention. More advanced strategies include dataset refining from novel data which closes a data flywheel loop where the more data passes through the system, the more robust and reliable it becomes.

4. Network-based Road Safety Assessment

Road safety is a topic of crucial importance for all EU member states and worldwide. Road safety is a major public health issue in the EU. Every year, thousands of people are killed or injured in road accidents. To reduce the number of road fatalities and injuries, it is important to identify and address the factors that contribute to road crashes [3].

The Network-Wide Assessment (NWA) methodology is a two-step process that assesses the safety of a road network using both reactive and proactive methods. The reactive method uses crash data to identify areas of high risk, while the proactive method uses road features to identify potential hazards. The two methods are then combined to produce a comprehensive safety rating for the road network.

The NWA methodology has been adopted by several countries in the EU, and it has the potential to make a significant contribution to road safety. By identifying areas of high road safety risk, the NWA methodology can help to prioritize road safety interventions and reduce the number of road fatalities and injuries.

4.1. Scope

The Network Wide Road Safety Assessment Methodology focuses on improving the safety of roads within the jurisdiction of the European Union, with a specific emphasis on the trans-European road network, motorways (both rural and urban), primary roads outside urban areas, and other roads that don't serve properties bordering on them but are built using Union funds.

Created under the guidelines of Directive 2008/96/EC, as later amended by Directive 2019/1936/EC, the methodology is targeted towards those roads that hold strategic, economic or mobility significance within the concept of European connectivity. This includes roads that are an essential part of intra-EU trade flow, mobility of EU citizens and supply chain pathways. [14]

The methodology optimizes road safety by focusing both on reactive (existing) and proactive (anticipated) roadway and behavioural variations. It addresses a wide array of parameters including segmentation criteria of road sections, estimation formula for crash density metrics, and different factors contributing to on-road safety in terms of design and operational characteristics. Some of these can include lane width, road curvature, road-side conditions, junctions, density of property access points and more.

Even though this methodology is designed for a defined set of roads, it isn't limited to them. It can be voluntarily applied by Member States to other roads situated outside urban areas. This ensures that Member States can utilize this tool for enhancing the safety of a broader spectrum of their roads network, thereby improving overall national road safety outcomes. This also accelerates the drive towards meeting the global standards of road safety and enhancing quality of transportation within the Union.

4.2. Reactive approach

The reactive methodology pertains to a systematic approach used in the Network Wide Road Safety Assessment. This method targets the evaluation of safety performance based on actual traffic accidents and crash data. [14]

The first step in this methodology is termed network segmentation. In this process, the entirety of the road network is divided into smaller and more manageable parts. These individual segments are formed according to certain criteria. These determinants could be related to road characteristics, geometry,

IOP Conf. Series: Materials Science and Engineering 1297 (2023) 012020 doi:10.1088/1757-899X/1297/1/012020

traffic volume, or a combination of these factors. The study mentions that three different approaches for network segmentation could be practiced. One approach deals with dividing the road into segments and junctions. The second method of segmentation includes road segments and junctions but with predefined dimensions considered. In the third approach, the sections are characterized by segments and junctions with measured dimensions.

In the subsequent step, data collection is conducted which involves the sourcing of crash data. This data is accumulated over a span of at least three years. Without this data, the implementation of the reactive methodology is not possible because it is data-driven if crash records over the said period are not available.

Building on the obtained data, the next step involves the calculation of safety performance metrics. This quantitative evaluation is derived from the accumulated crash data. The metrics are valuable as they give a numerical representation of the road's safety performance.

Once these metrics are calculated, thresholds of safety are estimated. These thresholds are established to classify road junctions and segments into different safety categories. This threshold-based classification plays an important role in understanding the safety conditions of different segments and devising interventions accordingly.

Finally, a safety ranking process is conducted. In this process, the road segments and junctions that were previously categorized according to safety performance and thresholds, are further ordered, or ranked. This step gives a clear hierarchy of safety performance for different segments. The primary goal of safety ranking is to create a clear and comprehensive line-up of segment-wise safety performance. Each segment or junction is allotted to a certain category of safety based on the statistical analysis of available crash data.

In essence, the reactive methodology is fundamentally discriminating the road type, whether it be an urban or rural motorway or divided rural road, etc. To ensure the effectiveness of this methodology, it requires detailed and high-quality crash data.

4.3. Proactive methodology

The proactive methodology described in the paper involves a detailed process that aims for a thorough safety assessment of roads. This method is considered proactive because it aims to forecast and prevent potential road crashes before they occur. The process starts with the identification of the road type that is to be assessed. The road could be an urban or rural motorway, a primary divided rural road, or a primary undivided rural road. By characterizing the road type, it allows for more tailored and specific measures to be implemented. Following road type identification, the network is segmented. [14]

Like the reactive methodology, the proactive method also necessitates the breakdown of the broad road network into smaller, more manageable sections. These sections can be based on the properties of the road such as the number of lanes, the existence of interchanges or junctions and their horizontal alignments. The segmentation here differs in criteria from that in the reactive methodology. Once the segments are determined, data collection is conducted. The required data includes various factors and parameters relating to the road, which potentially affect safety. Examples of parameters include the existence of interchanges or junctions, number of lanes and horizontal alignment, amongst other things. This collected data then undergoes an analysis process. Depending on the type of road under consideration, the relevant parameters are selected and weighed according to their impact on safety. The scoring system provides a quantitative measure of the safety risks of each segment.

Lastly, the scoring results are integrated and used for final ranking. This helps visualize and compare the proactive and reactive methodologies for the same section of the road. The outcomes of both methodologies inform decisions for interventions, prioritizing areas in need of improvement.

Overall, the proactive methodology allows for a data-driven, systematic, and predictive approach to improve road safety. By considering the inherent characteristics and conditions of the road sections and applying quantitative assessments, it provides a comprehensive view of the network's safety condition which in turn could inform effective preventive measures.

IOP Conf. Series: Materials Science and Engineering 1297 (2023) 012020 doi:10.1088/1757-899X/1297/1/012020

4.4. Integrated approach

The integrated approach is designed to synthesize the results of both the reactive and proactive methods into a comprehensive assessment system. The aim of this approach is to provide a holistic perspective that identifies risk areas across a road network to better prioritize future safety improvements. [14]

In the integrated approach, each road segment is assessed using both proactive and reactive methodologies. The results are then combined using a straightforward yet meticulously designed rule-based system. This system classifies each road segment into one of five classes.

These classes, each corresponding to a color and given a name, indicate the urgency or priority level of safety-related actions necessary after the assessment. They are as follows: - Class 5 (Worst performing): Very High Priority - colored red - Class 4: High Priority - colored orange - Class 3: Intermediate Priority - colored yellow - Class 2: Low Priority - colored light green - Class 1 (Best performing): Very Low Priority - colored dark green.

The goal of this classification is to make the final assessment easily understandable and actionable. Practically, this categorization helps road authorities to make an informed decision on the allocation of resources and deployment of measures for road safety improvements, particularly to areas labeled as high priority.

To visualize and compare outcomes of the proactive and the reactive methodologies for the same part of the road, one can use excel or similar tools with color coding. For example, segments scored differently on the proactive and reactive methodologies will visually stand out, serving as indicatives of the need for further detailed assessment and safety enhancement. By harmonizing the results of the reactive and proactive methodologies, the integrated approach presents a comprehensive view of road safety based on both historic accident data and the potential risk areas identified by the proactive methodology.

5. Application of AI and Deep Learning in NWRSA

The application of AI and deep learning in Network-Wide Road Safety Assessments provides several significant benefits. AI is efficient in analyzing and processing large volumes of data from various sources such as geographic information, photos, and documents. Another important capability of AI includes real-time recognition and classification of road assets like traffic signs, furniture, road surface deterioration, obstacles and overall inventory.

This reduces the need for time-consuming human intervention, making the process both cost-effective and efficient. With comprehensive safety evaluations, AI models take into account road infrastructure data, traffic data, and accidents, assisting competent authorities in identifying vulnerable areas and planning preventive measures. AI systems can continuously improve with dataset refinement from new data, becoming more robust and reliable over time. Furthermore, AI technologies enable tasks such as pavement condition evaluation and traffic flow analysis to be undertaken, leading to a deeper understanding of road conditions and traffic patterns. Altogether, these benefits contribute to more efficient management of road infrastructure, improved road safety, and data-driven decision-making.

6. Discussion

In the discussion of the use of artificial intelligence (AI) and deep learning for road network inventory and safety assessment, we inevitably confront differing perspectives. Critics of this method may express concerns regarding the reliance on technology, particularly on the aspects of data privacy and technological fallibility. The issue of data quantity is also a hinge point, as the accuracy of AI's evaluation is dependent on extensive and reliable data inputs, which might not always be feasible. Another notable disagreement may arise regarding the cost of introducing and maintaining such sophisticated systems, as well as the potential displacement of manual labor. It's essential to ponder whether the benefits of improved safety and efficiency offered by AI surpass these concerns. On the other side of the debate, advocates argue how AI can shape a safer and more efficient future in road management. But can we strike a balance between these two sides, perhaps with appropriate regulation and phased implementation? Furthermore, several questions still need exploration for optimal utilization of AI in this domain. For instance, how could we integrate AI-based methodologies in areas with limited access to high-quality data? How might we enhance AI's predictive capabilities to preemptively identify vulnerable locations and potential accidents? And how do we ensure the ethical handling of the data utilized by these systems? These queries pave the way for future research in this area.

7. Conclusion

In conclusion, this paper has explored the significant role of Artificial Intelligence (AI) in road network inventory and overall road safety assessment. AI models are capable of analysing a vast amount of data, rapidly identifying and classifying road assets in real-time, and providing comprehensive safety evaluations. This technology enables authorities to identify vulnerable areas, implement preventative measures, and plan strategic interventions, thereby contributing to a significant improvement in road safety. Furthermore, AI systems continually refine and improve over time, becoming more robust and reliable as more data is processed. These capabilities offer great potential for future advancements in road infrastructure management and safety assessments. Despite potential challenges with large-scale implementation, data privacy, and technological errors, the benefits highlighted in this study indicate that AI and deep learning will undoubtedly play a considerable part in shaping the future of road network and safety analysis. Further research is needed to maximise these benefits, overcome challenges, and realise the full potential of this emerging technology.

References

[1] Gandhi P, Khanna S, Ramaswamy S, 2016, Which Industries Are the Most Digital (and Why)?, *Harvard Business Review*

[2] Balali V, Armin R, Golparvar-Fard M, 2015, Detection, classification, and mapping of u.s. traffic signs using google street view images for roadway inventory managemen, *Journal of Visualization in Engineering*, 3, 1-18

[3] Popova V, 2022, Second Macedonian Road congess proceedings, Road Safety Assessment Practices and their implementation in Bulgaria

[4] Grozdanov H, 2015, External costs caused by road accidents, *TCI*, 9, 53-56

[5] Grozdanov H, 2019, Methods for estimating the external costs of road accidents, *National conference proceedings*, 44-56

[6] PIARC, *Asset management manual*

[7] Tao, V.C. Mobile mapping technology for road network data acquisition. *J. Geospat. Eng.* 2000, *2*, 1–14

[8] Sairam, N.; Nagarajan, S.; Ornitz, S. Development of Mobile Mapping System for 3D Road Asset Inventory. *Sensors* 2016, *16*, 367

[9] Elhashash, M.; Albanwan, H.; Qin, R. A Review of Mobile Mapping Systems: From Sensors to Applications. *Sensors* 2022, *22*, 4262

[10] Outay, F., Mengash, H. A., & Adnan, M., 2020, Applications of unmanned aerial vehicle (UAV) in road safety, traffic and highway infrastructure management: Recent advances and challenges. *Transportation research. Part A, Policy and practice*, 141, 116–129

[11] Kargah-Ostadi, N., Waqar, A., & Hanif, A., 2020, Automated real-time roadway asset inventory using artificial intelligence, *Transportation Research Part C: Emerging Technologies*, 116

[12] LeCun, Yann, Yoshua Bengio, and Geoffrey Hinton, 2015, *Deep learning*, 436-444

[13] Gagliardi V, Tosti F, Bianchini Ciampoli L, Battagliere ML, D'Amato L, Alani AM, Benedetto A. Satellite Remote Sensing and Non-Destructive Testing Methods for Transport Infrastructure Monitoring: Advances, Challenges and Perspectives. *Remote Sensing*. 2023; 15(2):418

[14] National Technical University of Athens, University of Zagreb, FRED Engineering s.r.I., *Network Wide Road Safety Assessment*, EC

IOP Conf. Series: Materials Science and Engineering 1297 (2023) 012021 doi:10.1088/1757-899X/1297/1/012021

Road safety impact assessment practices and recommendations

Vanina Popova

vkatzarova_fte@uacg.bg

Abstract. This article reviews the practices for conducting Road Safety Impact Assessment (RIA). An overview of current research and methodologies related to this field is presented, and various aspects of the preparation of RSIA are analysed.

The DIRECTIVE OF THE EUROPEAN PARLIAMENT AND OF THE COUNCIL on road infrastructure safety management (RISM Directive) defines the procedure "road safety impact assessment" as a strategic comparative analysis of the impact of a new road or a substantial modification to the existing network on the safety performance of the road network"; The RISM Directive provides only the basic framework for conducting such analysis by just pointing out its basic elements without providing specific methodology for it. Some countries have developed their own methodologies, others not.

Although the RIA is a procedure, that was included in the initial RISM Directive (2008), Bulgaria has not yet had much experience with it, nor has a developed methodology or recommendations for it.

The main aim of this article is to review and analyse international experience in conducting road safety impact assessment and to present some recommendations about it for such procedures from national teams.

The article draws attention to the methods and tools used in the preparation of the RIA. This includes the collection and analysis of road infrastructure and traffic data, the study of previous safety analyses, modelling and simulations of the proposed measures, and the assessment of social and economic aspects.

One of the main aspects discussed in the article is the possibility of using computer models and artificial intelligence (AI) in the preparation of the RIA. AI can be applied to process large volumes of data and perform complex risk and scenario analyses. This allows practitioners to obtain more accurate and predictable results to serve as a basis for making informed road safety decisions.

In conclusion, the article emphasizes the importance of preparing Road safety impact assessment and presents different practices and methodologies that can be used in this process. The use of computer models and AI can improve the efficiency and accuracy of RSIA and contribute to achieving higher standards of road safety.

1. Introduction

Road safety is indisputably one of the key topics to consider when it comes to planning and investing in road infrastructure projects. Traffic accidents lead to significant human and economic losses with inadmissible impact on the well-being of individuals and societies. Bearing in mind the importance of the proactive approach in the road safety management, it is crucial to implement possible impact measures as early in the project life cycle, as possible.

The "Road Safety Impact Assessment" (RSIA) procedure comes here as a helpful tool already in the inceptive stage of a road infrastructure project to identify the "safest" project alternative by pointing out specific safety benefits or deficiencies of given alternatives, also considering economic, environmental and other aspects too. Its main advantage is that it gives the opportunity to evaluate the potential safety implications of road projects and policy changes before they happen - before detailed planning and in help of the decision-making process.

The Directive of the European parliament and of the council on Road Infrastructure Safety Management 2019/1936/EU (RISM Directive) defines the procedure "road safety impact assessment" as a "strategic comparative analysis of the impact of a new road or a substantial modification to the existing network on the safety performance of the road network". As a result, the RSIA is expected to provide an evaluation of the safety outcomes of road projects and policy interventions, as well as substantiated data and recommendations for the project alternatives, that could and should be implemented further in the design process.

The purpose of the RSIA is to demonstrate, on a strategic level, the impact on road safety of different designing/planning alternatives of an infrastructure project. Road safety should play an important role when routes are being selected and safety awareness should be included in the decision making during the planning and design stage of the project. RSIA shall indicate the road safety considerations which contribute to the choice of the proposed solution. It shall further provide all relevant information necessary for a cost-benefit analysis of the different options assessed.[1]

Although the RSIA is a procedure, that was included in the initial RISM Directive from 2008 - 2008/96/EC, respectively transposed in the Bulgarian legal framework and requirements for road safety management, Bulgaria has not yet had much experience with it, nor has developed guidelines or recommendations for it. The RSIA remained as a mandatory procedure also in the amended RISM Directive from 2019 - 2019/1936/EU, however, it provides only the framework for conducting such analysis by just pointing out its basic elements without providing specific methodology for it. Some countries have developed their own guidelines, others not.

Therefore, this report focuses on this specific procedure, shedding light on its significance in promoting safer road environments and giving some common recommendations.

2. Problem definition and scope of the report

The main aim of this paper is to explore the scope of the RSIA procedure, review and analyse both national and international perspectives in conducting it and to present some recommendations and good practices.

The article draws attention to the methods and tools used in the preparation of the RSIA. This includes the collection and analysis of road infrastructure and traffic data, the study of previous safety analyses, modelling and simulations of the proposed measures or alternatives, and the assessment of social and economic aspects.

Considering the rapid development of artificial intelligence and related software and applications, the report also aims at identifying possible implementation of these in the different RSIA stages.

3. Road Safety Impact Assessment Practices

Countries across the globe have recognized the importance of the proactive approach in the road safety management, which in terms of RSIA is the safety evaluation of potential implications of road infrastructure projects before their implementation.

The procedure is often compared to the Environmental Impact Assessment and could be even carried out parallel to it. According to [2] the procedure costs normally less than 0,5 % of the total cost of a construction project, and when the cost of the construction project is more than 100 M. euro, then it could be about 0,2 % of it. Although these prices refer to period closer to year 2005, it should be marked, that prices could be much lower, especially for countries like Bulgaria.

IOP Conf. Series: Materials Science and Engineering 1297 (2023) 012021 doi:10.1088/1757-899X/1297/1/012021

As aforementioned, the RSIA's main objective is compare possible project alternatives and rank them from road safety point of view, taking into consideration the influence from and on the adjacent road network. Defining best alternative presumably excludes from further design the other ones, that go beyond acceptable risk levels, or the expected benefit isn't worth the investment.

With every project, there exists the potential for the existing conditions to be more favourable than any of the proposed alternatives. When a design team evaluates the preferred option, road safety impact is just one of the many factors taken into account. It is crucial to articulate the rationale behind the conclusions of the impact assessment, ensuring it receives appropriate considerations in the decision-making process. This approach aims to reduce the likelihood of future collisions resulting from planning choices or unintended consequences arising from road schemes designs. [3]

Although the basic concept of the Road safety impact assessment is the same in different countries, and the RISM Directive makes it mandatory for those, part of the EU (for roads part of the TEN-T Network and EU funded projects at least.), the specific implementation of this procedures varies within each country, which mostly depends on the availability of guidelines, previous experience and the use of software tools in each of the stages of the preparation of the RSIA.

The RSIA is an essential part of the planning process and is well established in some European countries, USA, Australia etc. United Kingdom, The Netherlands, Sweden and Germany are among those, who successfully integrate it into the planning and design of road infrastructure projects.

The Netherlands, for example, is one of the countries with greatest experience in this procedure with guidelines, developed already in 1994. – "Road Safety Impact Assessment: RIA". [4]

"Transport Infrastructure Ireland" has published their Guidelines in Oktober 2016. [3] , which set out the procedures required to implement Road Safety Impact Assessments on Infrastructure Projects affecting National Roads. They define the relevant schemes and stages in the design at which assessment shall be undertaken. The Process of conducting RSIA is described, including a methodology for it. For the details and procedures specific analysis required (Cost Benefit Analysis, for example), the document refers to an external document – the COBA Manual (Cost Benefit Analysis Manual).

Hungarian legal regulations mandate the completion of road safety impact assessments during the preliminary design stage, prior to making decisions about various alternatives. To provide more detailed guidance on the content of these assessments, the National Development Agency, responsible for EU-funded major infrastructure projects until 2013, commissioned the Hungarian Road Society in 2010. They were tasked with preparing a provisional guideline outlining the specifics of road safety impact assessments (National Development Agency 2010). From its initial publication in 2010 to 2015, this provisional guideline was applied in approximately 40 real-life cases in Hungary, accumulating practical experience. Distinguishing it from other guidelines, a smaller working group updated this document in 2015 (Universitas-Győr – TRENECON 2015). Further updates to this guideline occurred in 2017.[5]

It should be noted also that there are common international guidelines and best practices, provided by international organizations like World Health Organization (WHO) and PIARC (World Road Association), iRAP (International Road Assessment Programme) that are not explicitly dedicated to the procedure "Road Safety Impact Assessment", but could be successfully used in parts of such procedures, especially in the absence of national regulations and guidelines.

The "Safer Roads Investment Plans (SRIP)", developed by iRAP /EuroRAP for example, take into account various factors when evaluating road safety improvements. These considerations include existing road features, speed and volume of traffic, expected fatalities and injuries before and after treatment, the hierarchy of treatments, expected economic benefits of investing in that treatment, benefit to cost ratio and the internal rate of return. By considering these factors, the SRIP provides a comprehensive framework for prioritizing safety treatments on roads and highways. It helps decision-makers allocate resources efficiently to achieve the greatest possible reduction in road fatalities and injuries, while also considering the economic feasibility of these investments. [6] The SRIP involves other RAP protocols and tools and could be helpful especially when the RSIA isn't about a major project, but more about design alternatives of an existing road for choosing the best alternative, involving the

IOP Conf. Series: Materials Science and Engineering 1297 (2023) 012021 doi:10.1088/1757-899X/1297/1/012021

most efficient countermeasures from road safety point of view, that should be further implemented in the technical design stage.

4. Experience in Bulgaria

The procedure for RIA exists in the Bulgarian legal framework from year 2011 as part of „Ordinance No. RD-02-20-14/ 29.09.2011 on the scope and content of the assessment of the impact on road safety and of the road safety audit ". According to it, RSIA is carried out at the initial stage of planning before the approval of the infrastructure project. This gave a bit of wider time period and stages in the life cycle of a project, where such an analysis could be carried out and might also create some confusion in this respect.

With amendment of the RISM Directive in 2019, the respective Ordinance in Bulgaria had to be amended too, which happened in 2022 with the publication of a new "Ordinance on Road infrastructure Safety Management procedures (21.06.2022)". The only difference regarding the RIA procedure, which is to be noticed in the new ordinance is that the definition, that the procedure is carried out before starting feasibility studies or an investment project. The other thing is content, that is described in the appendix, where some differences could be noticed.

As both Ordinances are based on the requirements of the RISM Directives, all the procedures should mandatory be applied at least to the TEN-T network in the country and to road projects, funded by the EU. Road safety auditors in Bulgaria have already gained considerable experience in Road Safety Audits on different project stages (another procedure from the RISM Directive and respective Bulgarian Ordinances), though, since 2011, there are only a few RSIA procedures carried out in Bulgaria.

It turns out, that the situation is similar in other countries, because according to[7], by 2014 the RSIA is the least used procedure by member states.

The lack of experience also makes it difficult to assess expected positive effects of it on road safety on national or local level.

One of the main problems when conducting RSIA analysis in Bulgaria is the lack of systematic database. On the one side comprehensive historic accidents data and on the other economical evaluation of social costs of road crashes, as well as countermeasures costs and benefits.

A lot of countries, mostly upper-middle (Bulgaria), middle and low-income ones, rely on conventional road safety practices and often struggle to keep pace with the development of new technologies and innovative tools and approaches. In terms of RSIA, for example, this leads to the unexploited potential of learning from historical crash data on the one side, and of the proactive prediction modelling on the other one.

5. Scope and recommendations

RSIA is a systematic and iterative process, where various components are interconnected. Unlike most of the other road safety management procedures, the analysis in RSIA often extends above the pure road engineering matter and could require collaboration among transportation professionals, engineers, safety experts, economists and policymakers. To achieve optimal safety outcomes, it should be tailored to the specific characteristics of the road project and the regional context. The details may vary based on the project's scope and complexity, but there are some essential parts that should typically be included. Key components of one RSIA are the following:

- Project objectives and definition – introducing objectives and different alternatives, if given;
- Data collection and analysis – gathering comprehensive data related to the existing road and/or network of influence – road geometry, condition, traffic volumes, expected traffic growths, historical accident data and etc.;
- Identification of existing safety issues – a combination of the reactive and proactive approach
    a. Reactive – the historical accident data (including number of fatal and serious injuries, type of the collision and circumstances, geo-referenced localization and infrastructure

data) – for defining accident patterns over time and specific deficiencies of the infrastructure;
b. Proactive – Comparison of design alternatives upon technical parameters (if available) and potential risks – Risk assessment of the given/proposed project alternatives

- Definition of applicable countermeasures (when the project and project alternatives refer to an existing road; defining safety countermeasures and improvements to address the identified safety issues and risks);
- Project alternatives definition upon number and scope of possible countermeasures, including "Do-Nothing" option. (If not done earlier)
- Traffic flow forecast – modelling of future traffic growth by different alternatives and if possible, its influence on the safety;
- Quantification of risk and safety outcomes of different alternatives;
- Conducting Cost-Benefit Analysis, including other factors of influences (Environment, Social effects, etc.), comparison and ranking of different alternatives
- Impact Assessment Report

6. Application of artificial intelligence for Road safety impact assessment

Artificial intelligence (AI) is rapidly gaining power and applicability in our everyday and professional life. As technology is having an increasing role in road safety, the integration of AI and machine learning in RSIA in conjunction with human expertise and judgement could create a comprehensive approach to address road safety issues more efficiently.

AI – once used as Computer Vision (CV) for a quicker and more accurate acquiring of infrastructure asset and pavement inventory data and then secondly, for developing and running predictive models for the probability of crash occurrences and expected crash severity.

6.1. Data collection and analysis

The accurate and comprehensive data collection (accidents statistics, traffic flow, road inventory and pavement condition) is a foundation for effective road safety impact assessment, especially when it comes to alternatives to an existing road. The extraction and processing of such data by AI algorithms could help identifying patterns and trends for accident occurrences and to examine project alternatives based on similar correlations.

When it comes to initial safety assessment of existing roads, before further safety impact assessment of possible alternatives, the identification of safety issues is most often made based on onsite inspection or manually inspecting videos and images. Using Computer Vision (CV) instead could save time and be much more accurate.

CV is a field of artificial intelligence enabling computers to interpret and understand the visual world and to derive information from images, videos etc. There is already various research showing how computer vision algorithms have been developed for detecting roadside features, including guardrails, clear zones, rigid obstacles and roadside slopes. [8], but also, for pavement distress and condition assessment [9] and road network deterioration monitoring [10].

One of the challenges for road safety engineers and auditors when conducting road safety impact assessment is the lack of reliable data, being not enough information of the traffic flow, or about collision circumstances.

Extensive automated video data offers the opportunity for broader surrogate analysis, both in practical applications and research, serving the same objectives as the conventional historical accident data approach. These objectives include road network screening, countermeasure evaluation, road safety diagnosis, among others. Consequently, surrogate safety analysis is anticipated to serve a pivotal role as a supplementary safety approach, and in some cases, it has the potential to replace traditional methods, particularly when historical accident data is either scarce or unreliable due to its quality concerns.

This highlights the transformative potential of surrogate safety analysis, harnessing the wealth of real-time video data to enhance our understanding of road safety and inform proactive measures to reduce accidents and improve overall road safety outcomes.[11]

6.2. Accident prediction models

Machine learning (ML) is an area of AI that involves detecting patterns in data, making predictions or enabling actions to be taken without explicit programming in the form of the usual "if-then" routines and without classic automation and control engineering.[8]

Machine learning methods and algorithms have demonstrated their capability to create predictive models for various aspects of road safety, including accident risks, crash severity, and crash types. These predictive models can be invaluable in identifying more efficient countermeasures and evaluating project alternatives. [9] [10]

7. Conclusion

This paper provides a comprehensive overview of the international practices in conducting Road Safety Impact Assessment (RSIA), outlining its role in mitigating road accidents and enhancing road safety standards. Exploring different RSIA practices worldwide, it discovers that these are often diverse due to differing data availability and country-specific parameters.

The paper highlights the significance of leveraging Artificial Intelligence (AI) and computer models in RSIA. It identifies AI as a powerful tool in processing large volumes of data and conducting intricate risk and scenario analyses which can provide more precise and predictive results for informed decision making. One exciting prospect is the use of Computer Vision, a subset of AI, which can aid in efficient data collection and particularly provide a more accurate understanding of road conditions and potential pitfalls.

Furthermore, the RSIA process involves careful planning and integration of diverse disciplines. It requires a collaborative effort among transportation professionals, engineers, safety experts, economists, and policymakers for optimal safety outcomes. The paper emphasizes the necessity for these assessments to be tailored to the specifics of a road project and its regional context.

References

[1] Alfredas L, Vytautas G, Rasa Uspalyte-Vitkuniene, 2012 „Policy instruments for managing EU Road Safety targets: Road Safety Impact Assessment". *The Baltic Journal of Road and Bridge engineering.*

[2] Lacroix J, Gerlach J, 2008, „Sicherheitsmanagement für die Straßenverkehrsinfrastruktur - Neuer EU-Richtlinienvorschlag in der Diskussion". *Deutscher Verkehrssicherheitrat.*

[3] Road Safety Impact Assessment Guidelines, 2016, *Transport Infrastructure Ireland.*

[4] Road Safety Impact Assessment RIA, A proposal for tools and procedures for a RIA, 1994 : *SWOV Institute for Road Safety Research, The Netherlands.*

[5] Koren C, Mocsári T, Experiences with the implementation of the EU Directive on road infrastructure safety management in Hungary, 2018, *Proceedings of 7th Transport Research Arena TRA 2018, April 16-19, 2018, Vienna, Austria.*

[6] https://eurorap.org/investment-plans/. [Online]

[7] Ambros J, 2017 ,"Development of Czech road safety impact assessment guidelines",.

[8] "Automated safety assessment of rural roadways using computer vision" Report No. UT-23.06 , 2023, *Utah Department of Transportation, Research and Innovation Division.*

[9] Ayman H, Hakea EI, Fakhr M, "Recent computer vision applications for pavement distress and condition assessment", 2023, *Automation in Construction, Elsevier.*

[10] Parra N, Vargas-Calder´on V, Vinck-Posada H, Vinck N, 2022, "Road Network Deterioration Monitoring Using Aerial Images and Computer Vision", arXiv:2209.15455 [cs.CV]

[11] St-Aubin P, Saunier N, Miranda-Moreno L, 2015, „Large-scale automated proactive road safety

analysis using". *Transportation Research Part C: Emerging Technologies Volume 58, Part B, September 2015, Pages 363-379.*

Purpose of acceleration and deceleration lanes. Comparison of the regulatory framework for the design acceleration and deceleration lanes in Bulgaria, Austria, Germany and the USA (California).

eng. Marin Zhelev, Prof. Dr. eng. Valentin Nikolov

"Todor Kableshkov" Higher School of Transport, State, Bulgaria

mzhelev.ipc@gmail.com

Abstract. The report examines and analyzes the purpose of acceleration and deceleration lanes at two-level road interchanges. For the preparation of the current report are used the regulatory framework for the design of acceleration and deceleration lanes from Bulgaria, Austria, Germany and USA – California. The design elements of the acceleration and deceleration lanes are compared in detail. Widths and lengths of lanes, merging zones, angles and slopes. The conclusions of the report are given as guidelines for good practice in the design of road interchanges at one and two levels.

1. Purpose of the gateways

At the entrances and exits of road junctions, additional lanes are designed to the traffic lane of the highways and national roads for the inflow and outflow of motor vehicles to (from) the transit transport flow, also called acceleration or deceleration lanes.

The gateways are arranged at the expense of the forced stop strip or the bank of the road, providing a minimum width of the bank, according to the Road Design Norms[1].

Acceleration and deceleration lanes are used when crossing one and two-level intersections - road junctions. They allow motor vehicle drivers to speed up or slow down without affecting the movement of the traffic in the main direction.

Acceleration lanes allow vehicles entering the main direction to accelerate and reach the speed of the main flow. Slowdown lanes allow exiting vehicles to slow down to a safe speed without affecting the main flow of traffic.

At two-level road junctions, the deceleration and acceleration lanes may be connected, and the safety of the intertwining traffic flows is ensured by horizontal markings and vertical signaling.

With the implementation of acceleration and deceleration lanes, the average speed on the direct route is increased, delays are reduced and safety is increased by reducing the number of conflict points at slower and higher vehicle speeds.

2. Elements of the lanes

2.1. Deceleration lanes

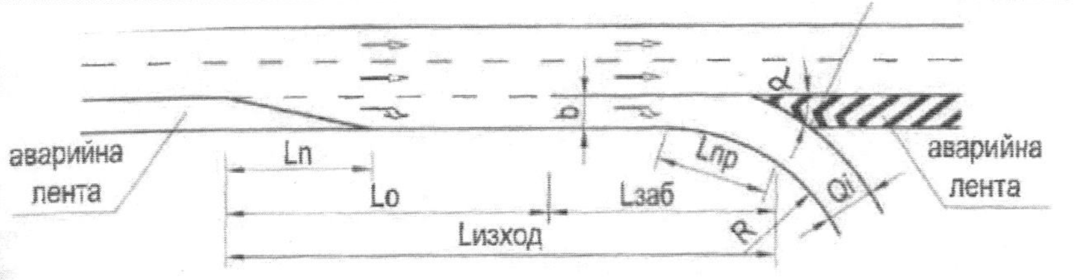

Figure 1. Scheme of a deceleration lane and its elements[2]

The length of the inflow lane or delay lane L_{Outlet} is formed by the following elements:

- $\mathbf{L_0}$ – length of the outflow section;
- $\mathbf{L_{dec}}$ – the section in which the speed of the motor vehicle is slowed down until it reaches a safe speed according to the restrictions;
- **Ln** – the transition zone for opening the outflow lane;
- **Lpr** – The section in which the transverse slope of the pavement is reshaped. The lanes receive the cross slope of the direct route. In this section, the slope of the pavement is reshaped from that of the direct route to the cross slope of the connection of the road junction depending on the radius of the connection, according to the NPP;
- **R** – radius of road link of the junction;
- **b** – width of the deceleration lane;
- **α** - angle of outflow.

2.2. Acceleration lanes

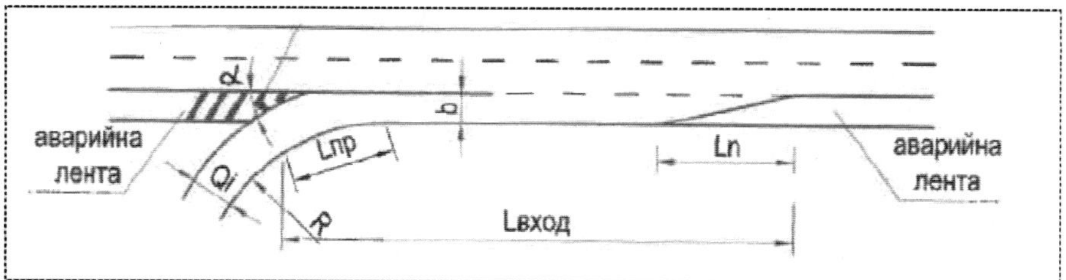

Figure 2. Schematic of an accelerator lane and its elements[2]

The length of a casting strip or accelerator gate L inlet is formed by the following elements:

- $\mathbf{L_n}$ – the transition zone for closing the inflow lane (acceleration lane);
- $\mathbf{L_{pr}}$ – The section in which the transverse slope of the pavement is reshaped. In this section, the slope of the pavement is reshaped from that of the connection of the road junction, depending on the radius of the connection, the NPP, to that of the direct route;
- **R** – radius of road link of the junction;
- **b** – width of the deceleration lane;
- α – angle of inflow,

3. Regulatory base

3.1. Normative base Bulgaria

3.1.1. Deceleration lanes
Design elements of deceleration lanes

Table 1. Length of the outflow lane L_0[2]

Road junction class	L0 [m]		
	L_0 recommended	L_0 min	L_0 exception
I class	200	170	150
II class	120	120	100

- $Lfec = \frac{V_0^2-V_R^2}{3.6^2*2*b_3}$ (m)[2]

IOP Conf. Series: Materials Science and Engineering 1297 (2023) 012022 doi:10.1088/1757-899X/1297/1/012022

where: V_0 – the outflow speed in km/h (80km/h for motorways and 70km/h for First class roads), V_R - the speed in a horizontal curve in km/h with a radius after the deceleration lane according to table 1, b_3– 2m/ s^2– acceleration while decelerating.

Table 2. Section length for Ldec [2]

R [m]	Motorway Ldec [m]	I class roads Ldec [m]
<40	110	80
<60	100	70
<80	90	60
<120	80	50
<150	70	40
<200	50	30
<250	30	-

Table 3. Speed in horizontal curve in km /h with radius - V_R [2]

Transverse slope Q %	R [m]	V_R [km/h]
q=6.00%	40	30
	50	35
	60	40
	80	45
q=4.00%	120	50
	150	55
	200	60
	250	70

3.1.2. Acceleration lanes

Design elements of the accelerator lanes

Table 4. Length on the section for inflow L in [2]

Road junction class	L_0 [m]		
	L_0 recommended	L_0 min	L_0 exception
I class	250	200	150
II class	250	200	150

3.2. Normative base Austria

3.2.1. Deceleration lanes

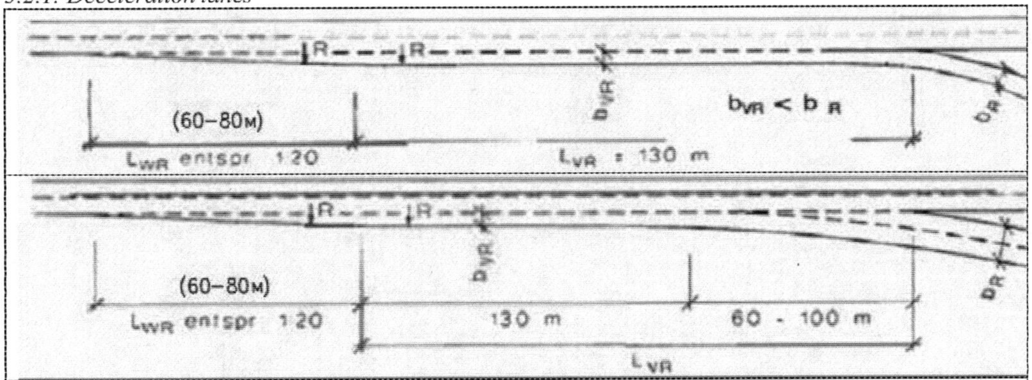

Figure 3. Schemes of deceleration lanes and their elements [3]

In the Austrian Standards [3], the transition zone for the opening of the outflow lane is specifically defined with an angle of 1:20 depending on the width of the lane.

3.2.2. Acceleration lanes

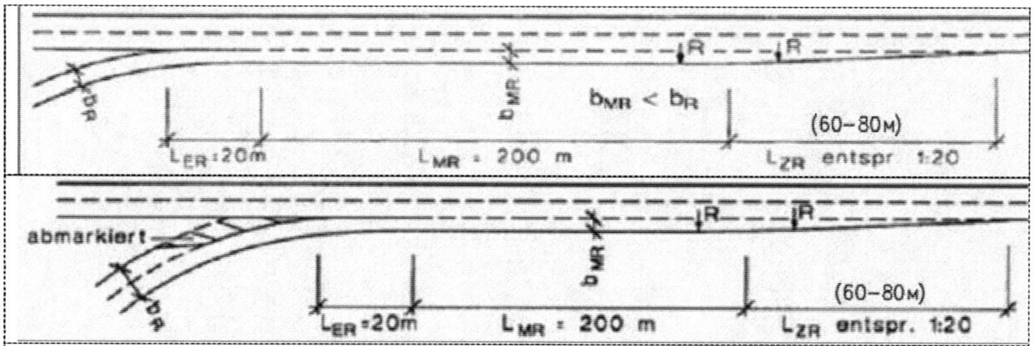

Figure 4. Schematics of an acceleration lanes and their elements[3]

In Austrian norms [3] the transitory zone for of the outflow lane is determined with angle 1:20 in dependency from the width on the lane.

3.3. Regulatory base Germany (RAA-2008)

3.3.1. Deceleration lanes

Table 5. The length of the dec. lane Loutlet [5]

Elements	Output type	PV 1st class	PV II class
L exit	Exit from DV	250	150
L p	For all outputs	60	30

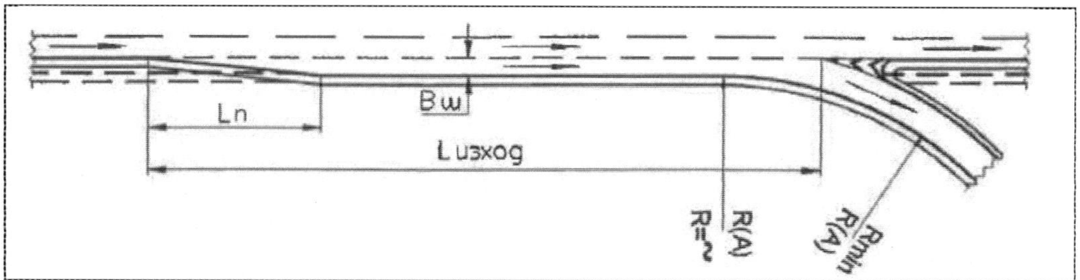

Figure 5. Scheme on deceleration lanes and their elements his [5]

3.3.2. Acceleration lanes

Table 6. The length of the inflow section Linlet[5]

Elements	Output type	PV 1st class	PV II class
L entrance	Entrance from DV	250	150
L p	For all inputs	60	30

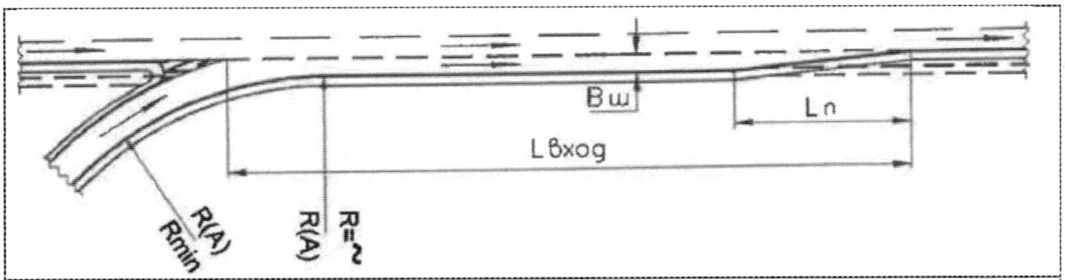

Figure 6. Scheme on an acceleration lanes and their elements [5]

3.4. Regulatory base USA (California)

3.4.1. Deceleration lanes

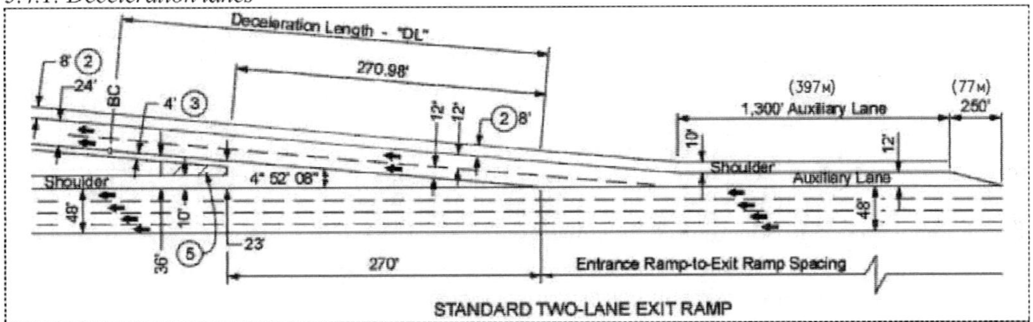

Figure 7. Scheme deceleration lane [4]

3.4.2. Acceleration lanes

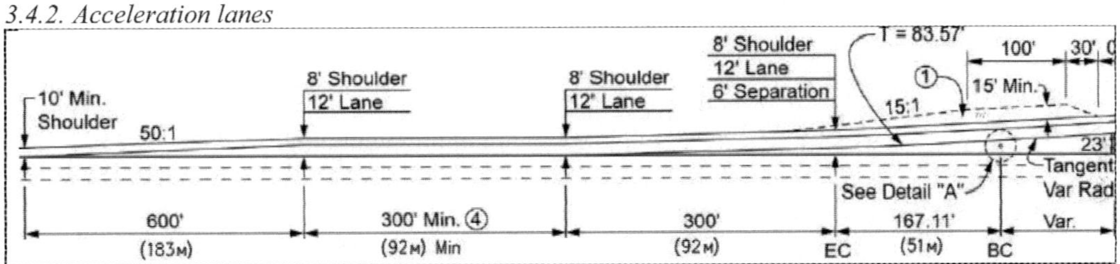

Figure 8. Scheme of acceleration lane [4]

Under the California Standards[4], the transition zone for closing the inflow lane (acceleration lane) is specifically defined with an angle of 1:50 or a length of 183m.

If the volume of trucks with 3 axles or more is greater than 5% of the participating vehicles to the highways, then the length of 92m can be extended.

3.5. Comparison of the lengths of the lanes in the considered normative bases

In the Bulgarian regulatory framework[2] the lengths of the acceleration and deceleration lanes are divided into two types for road junctions of the I and II classes. Three values for lengths Lrecommended, L min, Lexception are provided for them. For the comparison of the lengths of the lanes, the recommended length was used, with a minimum curve radius of 40m for the deceleration lanes.

In the regulations of Austria[3] and California[4], the values of the acceleration and deceleration lanes are precisely specified.

Table 7. Comparative table with the lengths of the lanes

Country	Lengths deceleration lane [m]		Lengths acceleration lane [m]	
	max	min	max	min
Bulgaria	310	290	250	200
Austria	210		300	
Germany	250		250	
California	474		367	

Conclusions:

- From the comparison it is clear that in the Bulgarian regulations[2] there are three types of varieties for the lengths of the acceleration and deceleration lanes, namely L recommended, L min, L exception, while in the Austrian, Californian and German regulations they are strictly defined.
- The lengths of the deceleration lanes in the Bulgarian regulations[2] are a function of the radius (R) as well as the cross slope (q).
- As can be seen from the comparison, the length of the acceleration lane in the Bulgarian and German legal bases [2] [5] are the shortest. But in the Bulgarian regulatory framework, its greatest value is adopted, while in the German one it is precisely specified.
- The lengths of the acceleration and deceleration lanes in the California regulatory framework[4] are the longest, which creates prerequisites for the safest inflow and outflow of traffic.
- When determining the lengths of the acceleration and deceleration lanes, the values in the Bulgarian regulations must be specified [2].
- The length of the acceleration lane in the Bulgarian legislation [2] needs to be specified, as it is used for intertwining high-speed traffic flows and is of key importance for safety in the area of road junctions.

References

[1] Nikolov V., Designing and construction of roads", Publishing base of VTU "T. Kableshkov", Sofia 2012, (ISBN -10: 954-12-0124-4).
[2] Ordinance No.RD-02-20-2 on road design , issue 9/10/11, 2018.
[3] RVS 03.05.13 - STRASSENPLANUNG, KNOTEN Gemischte und Planfreie Knoten, RVS 3.43, Seite 11
[4] HIGHWAY DESIGN MANUAL 500-1, CHAPTER 500 – TRAFFIC INTERCHANGES, May 20, 2022
[5] RICHTLINIEN FÜR DIE ANLAGE VON AUTOBAHNEN, Ausgabe 2008, (ISBN: 978-3-939715-51-1)

AUTHOR INDEX

Institute of Physics
Dirac House, Temple Back
Bristol BS1 6BE UK

ISSN: 1757-8981
ISBN 978-1-7138-8734-8

9 781713 887348

CURRAN ASSOCIATES INC.
proceedings.com